Changing Health Care Systems from Ethical, Economic, and Cross Cultural Perspectives

Changing Health Care Systems from Ethical, Economic, and Cross Cultural Perspectives

Edited by

Erich H. Loewy

and

Roberta Springer Loewy

University of California, Davis
Sacramento, California

Kluwer Academic / Plenum Publishers
New York, Boston, Dordrecht, London, Moscow

Library of Congress Cataloging-in-Publication Data

Changing health care systems from ethical, economic, and cross cultural perspectives/ edited by Erich H. Loewy and Roberta Springer Loewy.

p. cm.

Includes bibliographical references and index.

ISBN 0-306-46578-7

1. Medical ethics—Congresses. 2. Medical economics—Congresses. 3. Social medicine—Congresses. 4. Medical care—Social aspects—Congresses. 5. Medical care—Moral and ethical aspects—Congresses. I. Loewy, Erich H. II. Loewy, Roberta Springer.

R725.5 .C466 2001
362.1′042—dc21

2001029266

ISBN 0-306-46578-7

233 Spring Street, New York, New York 10013

http://www.wkap.nl/

10 9 8 7 6 5 4 3 2 1

A C.I.P. record for this book is available from the Library of Congress

Printed in the United States of America

This volume is gratefully dedicated to the Medical Alumni
Association of the University of California, Davis,
for their unwavering encouragement and support

Contributors

David Chinitz

David Chinitz received his BA with the Bennett Prize in Political Science from Columbia College in 1973 and his PhD in Public Policy Analysis from the University of Pennsylvania in 1981. He moved to Israel in 1981 where he has held positions as Coordinator of Social Sciences Research in the Ministry of Science and Researcher at the JDC/Brookdale Institute. He is currently Senior Health Lecturer in Health Policy and Management at the Hebrew University-Hadassah School of Public Health. Dr. Chinitz has also been a visiting scholar at New York University, Columbia University, and University of California, Berkeley. He has published articles and research reports and edited books on health care policy and regulation, consumer satisfaction with health system services, and priority setting in health care.

Larry R. Churchill

Larry R. Churchill is Professor of Social Medicine at the University of North Carolina at Chapel Hill and Co-Director of the newly established Center for Health Ethics and Policy. His major intellectual and research interests concern social justice in health care, the ethics of research with human subjects, and the moral dynamics of care at the end of life.

Thor Cornelius

Over the years, Thor Cornelius has had varied interests, from acquiring a degree in civil engineering and working for the California Environmental Protection Agency to studying medicine, philosophy, bioethics, and health care policy. He spent a year studying social philosophy and bioethics and is currently in his final year of medical school at the University of California, Davis. He plans to continue his medical training in the field of psychiatry while increasing his involvement in ethics and health care policy.

Faith T. Fitzgerald

Faith T. Fitzgerald is an Internist and Professor of Medicine and Assistant Dean of Students at the University of California, Davis, Sacramento. She took her MD at the University of California, San Francisco, in 1969, was board certified in Internal Medicine in 1973, and has been teaching medical students and residents ever since. She has written on a wide variety of topics in medicine, including protean disease states, medical education, physical diagnosis, and bioethics. She is a Master of the American College of Physicians, is currently Governor of the Northern California Chapter of the American College of Physicians, and sees patients in both hospital and clinic.

Michael J. Garland

Michael J. Garland, D. Sc. Rel., is Professor and Vice-Chairman in the Department of Public Health and Preventive Medicine and the Associate Director of the Center for Ethics in Health Care, Oregon Health Sciences University. He holds a doctorate in religious studies from the University of Strasbourg in France. Dr. Garland's research interests include ethical issues in the allocation of health care resources, social ethics education for medical students, ethics in human experimentation, and the community's role in guiding ethical choices in health policy. He cofounded Oregon Health Decisions in 1983 to foster public participation in the development of state health policy. The organization plays a continuing role in maintaining public involvement in critical policy choices affecting the Oregon Health Plan.

Walter Glannon

Walter Glannon is Assistant Professor in the Biomedical Ethics Unit, Faculty of Medicine, McGill University, and Clinical Ethicist at the Jewish General Hospital in Montreal. He is the editor of a forthcoming anthology of readings in biomedical ethics (Harcourt) and author of three forthcoming books: *Genes and Future People* (Westview); *The Mental Basis of Responsibility* (Ashgate); and *Fundamentals of Biomedical Ethics* (Oxford).

Reimer Gronemeyer

Reimer Gronemeyer was born in Hamburg in 1939 and studied theology and sociology in Hamburg, Heidelberg, and Bochum. He held an Assistant Professorship in Mainz before assuming the rank of Professor of Sociology at the University of Gießen (Germany), which he has held since 1975. His research has focused mainly on social politics and the sociology of developing countries. Professor Gronemeyer is currently working on several such research projects, one that focuses on the social impact of AIDS on society in South Africa, and one, with Professor Erich H. Loewy, that compares hospices in various industrialized countries.

Abraham (Avi) Israeli

Professor Abraham (Avi) Israeli, the Director-General of Hadassah Medical Organization, is Associate Professor of Health Care Management and Julien Rozan Professor of Family Medicine and Health Care at Hebrew University-Hadassah School of Medicine. Board certified in Internal Medicine and in Health Care Management, he holds an MD degree from Hebrew University-Hadassah School of Medicine, a Master's Degree in Business Administration from MIT Sloan School of Management, and a Master's Degree in Public Health from Hebrew University-Hadassah School of Public Health. He has published scientific essays in the areas of health care planning, administration and economics, and quality assessment of health care systems.

Rory Jaffe

Rory Jaffe, MD MBA, has been a physician for 20 years and recently received a degree in business administration. He is chief compliance officer for the UC Davis Health System and associate medical director of the UC Davis Medical Group. As associate medical director he was, in part, responsible for the development of the capitation distribution plan. As chief compliance officer, he is responsible for assuring that the health system abides by applicable government laws and regulations and is rendered less vulnerable to fraud, waste, or abuse. He is also responsible for ensuring that appropriate investigations are conducted into suspected improprieties.

Eike-Henner W. Kluge

Eike-Henner W. Kluge has a longstanding interest in different approaches to health care delivery and health care systems. While Director of Ethics and Legal Affairs of the Canadian Medical Association he counseled against the latter's challenge of the Canada Health Act. He teaches biomedical ethics at the University of Victoria, and acts as a consultant to various professional organizations and levels of government.

Erich H. Loewy

Professor Erich H. Loewy, while initially trained in medicine, has devoted most of his career to Bioethics with a particular emphasis on problems of social justice. In 1996 he was chosen, and continues to occupy, the first endowed medical alumni association Chair of Bioethics at the University of California, Davis. The author of several books and numerous book chapters and journal articles, Professor Loewy has taught extensively here and abroad.

Roberta Springer Loewy

Roberta Springer Loewy, PhD (Phil) currently volunteers as Assistant Clinical Professor in the Bioethics Program at the University of California, Davis. Having originally trained and practiced as a Critical-Care Nurse (1966-1982), she has since earned a BA in Liberal Arts from Skidmore College (1981), an MA in Philosophy with an emphasis on Health Care Ethics (1992), and a PhD in Philosophy and Ethics (1997) from Loyola University, Chicago. She has taught Bioethics in a number of settings and authored, co-authored, and edited a number of books, articles, and book chapters dealing with issues at the end of life, hospice, social justice, the nature of personhood, and the relationship between persons and society.

Jeanny K. Park

Dr. Jeanny K. Park is an Assistant Professor of Clinical Pediatrics at the University of California, Davis. She is a pediatric cardiac electrophysiologist who has also developed an interest in cross-cultural issues in medicine. Research interests include health services research into the disparities in morbidity and mortality of minority youth with chronic disease and the development and evaluation of cross-cultural issues in medicine curriculum.

Perry A. Pugno

Dr. Pugno has had training in both primary care and public health and is board certified in both family practice and emergency medicine. As a medical educator and clinician for more than 23 years, he has had extensive exposure to the diverse settings associated with outpatient and office-based care. His recent experience as an officer of a large, multi-state, integrated health care system in the most heavily penetrated managed care market in America has focused his concerns regarding managed care and the ethical challenges facing practicing physicians today.

Ben Rich

Ben Rich, JD, PhD, is an Associate Professor in the Bioethics Program at the University of California, Davis and visiting faculty at the University of California, Davis School of Law. Previously he was Assistant Professor and Assistant Director of the Program in Health Care Ethics, Humanities and Law at the University of Colorado Health Sciences Center and visiting faculty at the University of Colorado School of Law. He has published articles on a wide range of topics in biomedical ethics and the law, and is currently working on a book that explores the role of law in medical ethics and medical practice.

Susan E. Zinner-Kemp

Susan E. Zinner-Kemp received her master's degree in health administration and law degree from Washington University in St. Louis in 1992. She completed a post-graduate fellowship at the Hines Veterans Hospital in Maywood, Illinois, in 1992-1993. She joined Cook County Hospital and worked as an administrator in both pediatrics and medicine from 1993 to 1998. She left to join the full-time faculty at Indiana University's School of Public and Environmental Affairs (Gary campus), where she teaches health law, medical ethics, administrative ethics, and other health policy courses.

Preface

This volume is the result of a conference sponsored by the Medical Alumni Association of the University of California, Davis and held in Sacramento, California, in January, 2000. The purpose of this conference was to examine the impact of various health care structures on the ability of health care professionals to practice in an ethically acceptable manner.

One of the ground assumptions made is that ethical practice in medicine and its related fields is difficult in a setting that pays only lip service to ethical principles. The limits of ethical possibility are created by the system within which health care professionals must practice. When, for example, ethical practice necessitates—as it generally does—that health care professionals spend sufficient time to come to know and understand their patients' goals and values but the system mandates that only a short time be spent with each patient, ethical practice is made virtually impossible. One of our chief frustrations in teaching health care ethics at medical colleges is that we essentially teach students to do something they are most likely to find impossible to do: that is, get to know and appreciate their patients' goals and values. There are other ways in which systems alter ethical possibilities. In a system in which patients have a different physician outside the hospital than they will inside, ethical problems have a different shape than if the treating physician is the same person. This is not to say that one system is better and the other worse; it is to say that each has its own problems and therefore approaches will differ.

Another of the ground assumptions is that the creation of a just system or institution is difficult unless the culture and society attempting to establish such a system meet criteria for being a just society. This is one of the true problems faced in the United States today. We have nearly 45 million uninsured whose access to health care (unless they are critically ill) is through capricious charity. But health care is not the only or even, we shall argue, the most important lack in our society. The gap in income between the opulently wealthy and the grindingly poor is wider than it is anywhere else in the industrialized world, and it is growing. The minimum wage paid today is insufficient to keep people above the poverty level despite the fact that the poverty level itself is already set at an unrealistically low point! A large number of persons are hungry and many are homeless. This basic injustice in a thriving economy is not something that we can or should watch with equanimity. Thus health care ethics as usually understood (that is, the branch of health care ethics which pertains to problems at the bedside) in the United States turns out to be "rich man's" ethics—that is, it deals predominantly with the concerns those who have access find compelling. Those without access are not concerned with problems of limiting care at the end of life or with informed consent—problems that concern those of us who are being treated. Those without access are concerned with receiving treatment in the first place. Furthermore, getting treatment when ill is really a secondary concern to those who lack the everyday necessities to sustain acceptable life—as Brecht so well put it, "Erst kommt das Fressen, dann kommt die Moral" ("first comes eating and only then comes morals").

We hope that the papers we have selected from this conference (both those papers given by invited speakers and those submitted papers chosen from the parallel sessions) will remind us of the diversity of approaches available to us and help to illustrate the problems these approaches may entail.

Erich H. Loewy
Professor and Alumni Association Endowed Chair of Bioethics
University of California, Davis

Roberta Springer Loewy
Assistant Clinical Professor, Bioethics
University of California, Davis

Contents

Changing Health Care Systems from Ethical, Economic, and Cross Cultural Perspectives

Chapter 1

Health Care Systems and Ethics

Erich H. Loewy
Professor and Endowed Alumni Association Chair of Bioethics
Associate Professor, Philosophy
University of California, Davis
Sacramento, California 95817
e-mail: ehloewy@ucdavis.edu

Key words: health care systems, single-tiered, multi-tiered, single-payer and multi-payer systems; "poor law" and "welfare" approaches; "rich-man's" ethics, social justice, democracy, socialism, communism, communitarianism, capitalism, marketplace, managed care, John Dewey

Abstract: This paper gives a brief and basic introduction to some of the concepts and vocabulary used in the debates about health-care systems. It differentiates between socialism and communism and points out that the two are hardly identical and that democracy and capitalism are not necessarily related. The difference between single- and multiple-tiered health care systems and arguments for and against each are briefly discussed. The attempt to deal with our ethical problems in health-care and to create a just health-care system may by itself positively affect what is now perceived to be a basically unjust society.

1. INTRODUCTION

Physicians are confronted with an ever-increasing number of ethical problems. Some of these problems are old problems that have been complicated by the ever-increasing technical ability of medicine; others are new problems brought about by entirely novel and unanticipated technologies;

Changing Health Care Systems from Ethical, Economic, and Cross Cultural Perspectives,
edited by Loewy and Loewy. Kluwer Academic/Plenum Publishers, New York, 2001.

still others are ethical problems which have been brought about by economic factors and a changing health care system.

The problems of ethics at the bedside practice of medicine in the United States today are what I have previously referred to as "rich man's" ethics. That is, they are the problems which concern those of us with ready access to medical care.[1] Questions of terminating treatment, of futility or of in vitro fertilization are problems of little concern to those forty-two to forty-five million of us who cannot have access to preventive or early curative treatment. Our attention in health care ethics has been predominantly on those bedside problems that affect the insured. Although ethicists have occasionally paid lip service to equitable access for all, they have refused to take an organized, let alone an effective, stand. Neither has organized medicine invested a great deal of energy in pursuing an agenda they profess to embrace. It is, I would claim, most difficult if not indeed impossible to practice ethically within an unethical system just as it is difficult to create a just system within a basically unjust society.

Even when there is equitable access, the nature of the system shapes the ethical problems that physicians confront and limits the responses they can have. Thus a system with free choice of physicians or a system in which patients are seen by different physicians inside or outside the hospital will affect the relationship that physicians, patients and other health care professionals have with one another. Moreover, such a system will, therefore and inevitably, shape how and what we come to recognize as ethical problems. I am not arguing for one or another system—I am merely stating that to understand and truly appreciate the problems and the options one must understand the system. When systems stand in the way of ethical practice, physicians, I shall argue, have two obligations:

1. To do the best they can within the system (which, at times, may even include "gaming the system" as perhaps the lesser of several evils)
2. To play an active part in changing the system itself

For the sake of this discussion, I shall assume but not argue that a decent community that can afford it, is obligated to supply at least basic health care for all within its borders.[2] Such an assumption is grounded in an understanding of what defines a well functioning and decent community—one which tries all it can not to disparage its members.[3] A libertarian model may provide structure for a loose association of people united by obligations of non-interference with one another and an adherence to freely entered contract but such a model will fail to yield the solidarity communities require if they are to flourish and evolve. When individuals recognize that their individual goals can be pursued with a good chance of success for all only within the embrace of a community, and when society is aware that its success depends upon fostering individual skills and talents, communities will have a solidarity based on their mutual and intertwined common goals and ends. A community that unnecessarily leaves persons uneducated, bereft of the basic necessities of life or without equitable access to health care plays a role in disparaging some members at the expense of others. Communities in which individuals feel a strong bond with one another are communities that will prosper, evolve and endure—such communities will accept the obligation to meet basic needs as a condition of successfully association.[4]

The most erudite discussions of the finer details of justice or the professions of despair by the medical community at the number of uninsured are pointless without political action. This is not a new observation: Aristotle long ago saw politics and ethics as firmly entwined. Questions of ethics are questions directed at courses of action—action which when it comes to systems can only be modified within a political context. It is my thesis that those persons associated or concerned with the ethical practice of medicine have an obligation to take an active role in creating a system in which ethical practice can take place. Such an obligation transcends that of the ordinary citizen. It is one which (and with particular force) ethicists who are supposedly the most concerned about ethical practice should eagerly embrace. Doing one's job as well as one can—or teaching the finer points of ethical theory—is pointless if the constraints of the system force one to practice in a way which one readily recognizes as being ethically problematic.

The health care system in the United States is the most expensive, the most inequitable and the most bureaucratized in the world. As good as the care of critically ill patients still is in the United States, even that is no longer the best there is. We in the United States today have become very skilled at remedying crises we could have easily prevented. Often we remedy an acute crisis only to send patients out into the very same situations that produced the crisis in the first place. Not only is this ethically problematic but it is, in the long run, economically unwise.

Most people recognize that the various solutions proposed for remedying the problems of our health care system have not only *not* turned out to be solutions but have, in fact, made the problem worse. Managed care (especially for-profit managed care) which was to be the American answer to what is improperly called "socialized medicine" has resulted in even more people being uninsured and without proper access. Managed care, as I shall show, has made the ethical practice of medicine most difficult, has distorted the patient-physician relationship and has added a new layer of ethical problems without solving the old. Most of us would agree that an equitable health care system in the United States is sorely needed.

2. BASIC LANGUAGE AND CONCEPTS

A system is something that has some sort of internal coherence and controlling elements. The cardiovascular system and the educational system are examples. In the health care system as it exists in the United States today the only internal coherence and the only controlling element is a theory of the free market—and not even that is entirely carried into practice.

Before we can speak of building a health care system, certain basic concepts, terms and language must be agreed upon. Many terms are loosely used and need to be defined. What follows is an attempt to define some of these terms.

2.1 Economic and Political Distinctions

The term "socialized," since it is bandied about rather freely, must be understood. Socialism, first of all, is a term often equated with communism. This is untrue and inaccurate. Communism denies the right to private property; socialism recognizes the right to private property but insists that the fruits of labor ought, by right, go to those who work. Thus, worker ownership of United Airlines or the Saturn Car Company is, in a sense, a form of socialism. Furthermore, socialism importantly holds that certain goods and institutions essential to the community should be owned and controlled by it (and this is where the term "socialized medicine" comes in). Persons here in the United States are persistently taught that democracy necessarily entails capitalism and that capitalism furthers democracy. Nothing could be further from the truth. First of all capitalism is an economic system and socialism a political system—and while economic and political systems should preferably fit together, they are not synonymous. Both capitalism and socialism can exist in a monarchy, a dictatorship or a political democracy. The philosophical basis of capitalism is the freestanding, largely asocial individual, whereas the philosophical basis of socialism as well as of democracy is community. Social democracy is a democracy that emphasizes democratic process and accepts social responsibility; democratic socialism is a system in which the means of production are predominantly in the hands of those who have a part in creating the product and in which decisions are made in a democratic fashion. In democratic socialism, private capital exists but is strictly regulated and the community controls many things basic to communal life (things like health care, education and public utilities).

Most national health care systems are not "socialized"—that is, they are not operated by the state. A socialized system is one in which the state from general taxation creates, maintains and operates a health care system. Many if not most systems in the industrialized world that provide at least basic health care to all citizens do not meet such a definition. They are operated by and through various usually government supervised insurance schemes; but they are not, in the true meaning of the word, socialized.

If we are to allow the market to control the distribution of a commodity, we must first ask if the philosophy of the market is appropriate to the particular commodity. The basic philosophy of the market presupposes that consumers have sufficient:

1. Resources to participate in the market
2. Understanding of what constitutes a "good" product for them to choose
3. Leisure to "shop around" and compare quality and price
4. Protections against fatal injury, should they make a "wrong" choice

In health care none of this applies.[5] In general, when it comes to health care, consumers do not have funds sufficient to engage in a free market. They do not and cannot understand what a good product would be. They have little time when ill to "shop around and compare;" and, should they make a wrong choice, they might well be fatally affected. Beyond this, the philosophy of the market requires that consumers and purchasers are one and the same entity: they can weigh their personal idea of price and quality and, within the limits of their financial possibility, come up with a decision reflecting their evaluation. In the United States today the consumer (the patient) and the purchaser (almost invariably the patient's employer) are interested in quite different things—the patient in quality and accessibility, the purchaser in cost.

2.2 Single and Multiple, Tiered and Payer Distinctions

If one is thinking of creating a health care system one first of all must decide whether such a system should be single- or multiple-tiered and who should pay. Although the two terms are often used as though they were synonymous, a single payer system is not synonymous with a single-tiered system. In a single payer system there is one agency (be it government or private) which pays out "benefits." This payer could conceivably be a large insurance company selling different policies to different persons: i.e., one payer who pays differently for different persons.

In a single-tiered system all get the same of a given commodity and no one can buy more; in a multiple-tiered system a basic minimum is provided and more can be bought by those willing or able to buy more. All getting the same could mean that all insured by a given company (let us say all members of Kaiser-Permanente) get the same services or it could mean that all members of a community under the umbrella of a communal plan receive the same services. In the way I shall use the term I am referring to all members of the community. In most communities and as a general rule, fire and police protection are single-tiered, while education is multiple-tiered. In terms of health care, all "getting the same" refers to those things which affect outcome: physicians, nurses, waiting time, procedures, drugs, etc. A single-tiered system, the way the term is used here, might well be one in which the affluent could purchase a private room, nice curtains on the windows and a bottle of wine with dinner. But they (the affluent) could not buy different physicians, a shorter waiting time or a hospital bed that is better staffed than another. Multiple-tiered systems provide a basic minimum to all and leave additional services up to the individual's ability and desire to buy them. Canada and the Scandinavian systems are essentially single-tiered systems whereas the British system is multiple-tiered. A single-tiered as well as a multiple-tiered system can be socialized or not. It is conceivable that the state could manage and finance a system in which one class of employee would receive different benefits from another and it is equally conceivable that a system operated through insurance companies might be essentially single-tiered.

3. A BRIEF COMPARISON OF SYSTEMS AND THEIR ETHICAL IMPLICATIONS

Various countries have adopted a variety of health care systems. In all of these countries there is one common denominator—they all provide at least basic health care coverage to virtually all residing within their borders. The United States, as has been said, is unique in *not* doing this.

The Scandinavian countries differ among themselves but have two important features in common: they are exclusively publicly funded and they use primary care physicians as gatekeepers. Germany, Austria and to some extent Switzerland are funded via mandatory employer/employee contributions, have a strictly regulated (but becoming increasingly less strictly regulated) insurance system and provide insurance for those who would be otherwise uninsured. The United Kingdom has a multiple-tiered, nationalized system with the national health care sector publicly funded. Canada's system is single-tiered and nationalized with public funds distributed among the provinces. France has a mixed system.[6]

Different systems spawn different ethical problems. A system in which physicians care for their patients both inside and outside the hospital (as is generally the rule in the United States) has somewhat different or at least differently shaped ethical problems than does a system in which ambulatory and in-hospital care are strictly separated. A capitated system offers different incentives than does one that is fee-for-service. Physicians who must deal with private insurance companies face different ethical problems than do physicians who are paid directly by the government.

In most systems a common denominator remains: physicians are primarily obligated to the good of their individual patients. That physicians are primarily obligated to the individual patient is a medical tradition as old as medicine itself. Furthermore, it is a tradition of medicine, which is and has, in a variety of cultures, been a constant societal expectation. Of course, expectations by themselves do not create obligations. But when expectations are consistently met over a long period of time, they become a justified expectation and eventually an obligation is created. In the United States and under our current system of Managed Care, this ancient tradition has frayed; the expectation that physicians are, above all, committed to their particular patients' good come what may, is often not met. Trust is attenuated and suspicion of the medical profession, unfortunately often not unjustified, escalates. Physicians today are often forced to choose between their patients' good, loyalty to their organization and personal self-interest. Increasingly physicians regard themselves more as good employees of their organization than they do as advocates for and of their patients.

The language we use conditions the way we think and often determines the way we feel and act. In the last few decades there has been a gradual shift in language, which both reflects and has driven these other changes in physician-patient relationships. Physicians have become providers; patients have become first clients, then consumers and now, even worse, customers. Often they are, in insurance jargon, simply spoken of as "lives." This shift in language (one still fairly unique to the United States) is, in my view, by no means accidental or trivial—it is a shift at the very least encouraged by those who stand to gain by the disruption of an ancient relationship. It is one that health care professionals buy into at their peril.

4. APPROACHES TO PROBLEMS

Whether we build a single- or a multiple-tiered system is in part dependent upon how we see ourselves related as individuals to one another and to our community. If we envision ourselves as united merely or mainly by obligations of mutual non-harm but by few if any obligations of mutual help, we will build a far different social system than if we see ourselves united not only by the obligation not to harm but equally by the obligation to help one another. There are two possible approaches: one is termed the "poor law" philosophy. A poor law approach is one in which a certain segment of the population is entitled to certain benefits if and only if they meet definite criteria—food stamps below a certain level of income might be one example. The other approach is what has been termed the "welfare" approach. In such an approach persons are entitled to certain basic goods and services not because they meet certain criteria, but by virtue of being members of the community—police and fire protection would be an example.[7] Most societies adopt a mixture of both philosophies–which predominates is a function of how we see ourselves related to one another and to our community. Societies that are more committed to accepting obligations of mutual assistance rather than merely obligations of mutual non-harm are more apt to construct single- rather than multiple-tiered systems.

There are sound arguments for both kinds of systems. Those committed to a multiple-tiered system providing at least a basic minimum to all, argue that persons ought to be free to support whatever values are most important to them. Persons who prefer to have a luxury car or an elaborate vacation trip should be free to make such a choice at the expense of more sophisticated health care. Further, people committed to a multiple-tiered system will argue that it seems unfair that persons who have worked hard and saved money should subsidize those who have either been lazy or profligate. A multiple-tiered system would give basic health care to all but reserve more sophisticated and more expensive care for those willing (out of pocket or through insurance) to buy such care. An argument about not caring for diseases that are the product of personal risk-taking is often appended to such an approach: persons who chose to live a healthy life style should not be asked to bail out those who smoked, drank or went bunjie-jumping.

Those who prefer a single-tiered system will argue on two levels: first of all, they will argue that a true community is properly committed to support the weak and frail. In such an argument support of everyone within the limits of a community's capacity is part of the definition of a true community. Communities that support their weak and frail (something we potentially all are) will, it is argued, show more solidarity, have a better chance to endure and, ultimately, offer their members a greater possibility for optimizing their values and pursuing their interests.

Second, those who prefer a single-tiered system will argue by countering the arguments that are made for a multiple-tiered system. The argument that all persons should be free to support their own values is true only within the context of a community that allows certain values to be expressed. We generally do not value our necessities until we are deprived of them: few of us give much thought to food or drink until we are hungry and thirsty—and if we failed to take enough money along to buy food and drink we shall go hungry and thirsty. Similarly, persons do not value their health until it is threatened. Unless they have "taken along enough money" it is quite possible that they will find themselves without access to medical care when it is most needed. Since few of us would wish to live in a community which allowed persons to die simply because they lacked foresight, were lazy or lived above their means (and fewer still would wish to see their families treated in this way), we would be likely to end up either paying for such care as a community or collecting private money. And, indeed, this is what frequently happens today. Uninsured persons are taken into charity hospitals or supported out of the public coffer when they become critically ill. Often private collections are taken up for those who are uninsured (and, therefore, not acceptable to most transplant programs). Here a weeping, well-dressed and soft-spoken grandmother, psychologically, "has it all over" someone sloppily dressed and using coarse language. Yet, upon critical reflection, such a state of events should accord with few persons' sense of fairness.

More importantly: people who fail to buy additional insurance are unlikely to be those who have been lazy or who have preferred to buy luxury items. People who fail to buy additional insurance are frequently not choosing between expensive automobiles and additional insurance but between additional insurance or food (or perhaps schooling) for their children. They have most often been neither lazy nor profligate: they have simply been hard working, poor and unlucky. Arguing that those who jeopardized their own health should not burden the community with the expense of treating the result of their behavior is not an argument against a single-tiered system. It is entirely possible to tax many of these activities and to use the tax revenue to support the additional health care. Whether or not this is fair is another question—but it most certainly can be done.[8]

Resources are limited. What is spent for one thing cannot be spent for another. Economists refer to this phenomenon in terms of "opportunity costs"—spending on one thing precludes the opportunity of spending the same money on another. Health care—important as it is—is neither the only nor the most important of several social goods.

Imagine the following experiment: Persons are asked to choose two from among three social goods. The one not chosen will be something that they would have to obtain by whatever private means they could; the ones chosen would be things guaranteed for life. The choices are: (1) having all biological needs met–the person will never go hungry, without shelter and so forth; (2) having all educational needs fully met; and (3) having complete and free access to medical care. The choice must be made behind a Rawlsian veil of ignorance. That is, the choosers will not know their age, sex, social standing, wealth or state of health.[9] Most prudent choosers would undoubtedly choose to have those things necessary to sustain life vouchsafed to them—after all, if one is not alive, nothing else means very much. Furthermore, most of us would choose full educational opportunity, for without it the content of our lives would be impoverished. With our biological necessities guaranteed and our educational needs met we would probably be able to gain access to medical care should we become ill. I do not argue that health care is unimportant or that a decent society should not in justice supply medical care to all—indeed, I feel that a community able to supply all three *is obligated to do so.* I do argue that important social goods must balance one another—as in a symphony one instrument cannot be allowed to drown out all others, so in a community one social good cannot be allowed to swamp all others.

Since resources are limited and demand is great, a system of equitable distribution is essential. To deny this fact is to delude oneself. Two steps are inevitable–the first is rationalization: that is, to expend our resources only for those things that are of accepted value, to stop waste and to eliminate duplication. Depending how these things are defined and done, few would argue against such measures. The problem, of course, is that what is and what is not valuable or wasteful is hardly self-evident. The second step is rationing, something that we have done for a long time and in all systems but have never really admitted to doing in any of them. In the United States we ration by ability to pay—those who (by insurance, out of pocket or through charitable funds) are able to pay receive services; others do not. With managed care some rationing—though not called by that name—also occurs by interposing a great deal of administrative work between request and fulfillment. This has been called the hassle factor and, although not ever called a rationing measure, surely works as such. It sharply reduces request for services—the greater the hassle, the more the chance that people will forgo what they had wanted. In other systems other ways of rationing (generally referred to by other names) takes place. I do not oppose rationing—I fear that it is inevitable. I *do* oppose not dealing with people in a straightforward and honest manner—if rationing is needed, call it that and defend it.

When physicians deal with their patient's problems, they are dealing with identified lives—that is, with persons they directly know or can identify as real persons. When we deal with people we recognize as persons and especially when we deal with such people in a setting of illness or misfortune our natural empathy is aroused. If, however, we are to help such people we need to engage more than our emotions—our emotions alone could lead us to do very destructive or omit very important actions. We need to temper our emotions with reason—ending up with what I have called "rational compassion." When rationing resources or building a health care system we deal with people we do not know and of whom we have no direct knowledge. Such unidentified or statistical lives engage our reason. But reason alone is cold and distant when it comes to dealing with human problems. In building health care systems or rationing resources we need to be mindful that such lives are neither merely statistical nor unidentified but merely not identified by ourselves. By virtue of being lives and by virtue of all human lives occurring in a social nexus, such lives are very much real and very much identified by others. It behooves us to try and visualize decisions we make in human terms—that is, to allow our compassion to help us understand what the lives of those for whom decisions are made are like. We need to season our reason with compassion—a step I have called "compassionate rationality."[10]

Tempering compassion with reason or seasoning reason with compassion necessitates the use of curiosity and imagination—human capacities that in our civilization tend to be downgraded. When dealing with identified lives curiosity prompts us to ask how certain courses of action would affect the person we are dealing with and imagination would allow us to sketch an answer; reason and compassion (but not compassion alone) would then allow us to make a choice. In dealing with statistical lives curiosity would impel us to ask what being in their shoes might be like and imagination would help us to achieve at least some understanding of their plight. Compassion together with reason but not reason alone would then allow us to come to a decision.[11]

Our concern with how to create a health care system or how to reshape one that exists is perhaps not the first concern we should have. Invariably when we are confronted with a problem in ethics we ask, "what shall we do." This is true no less in building a system or creating a policy than it is in facing problems at the bedside. This question, however, is not the first question we need to ask. The first question, I think, is not what shall we do but who is entitled to decide what should be done and then how the voices of those who should be involved in deciding should be heard. We have, I think, for all too long crafted an ethic *for* others—*for* the weak, *for* the sick, *for* the poor; it is time, I think, to craft an ethic *with* instead of *for* people. The creation of an ethic for others is a remnant of autocracy and monarchy that in a democratic society has outlived its usefulness. Clearly, if we truly believe in democracy, all those potentially or actually affected by a policy ought to share in shaping it.

Creating a health care system is a most difficult task. It is one in which physicians, nurses, economists, sociologists, ethicists, administrators and many others must contribute their expertise and work together towards a common solution. There is no doubt that none of us has sufficient expertise to come up with more than a small part of an answer. Nevertheless, the broad outlines of a policy are things in which the electorate at large should have a powerful voice. Issues such as whether a single- or multiple-tiered system would serve us best or what should and what should not be considered as necessary health care services are issues which concern the man or woman on the street who is ultimately affected. It seems self-evident that he or she should have an opportunity to have his or her voice heard.

Finally, one cannot create a just system in the context of an unjust society. Having all of those who are potentially or actually affected participate in crafting a solution entails a truly democratic system and not merely a *pro forma* political democracy in which everyone has a right to vote and in which counting of votes is at least not overtly fraudulent. A political democratic process—which ultimately seems to be the only acceptable way of creating public policy—necessitates that the preconditions of democracy are met. John Dewey long ago stated that at least three preconditions must be in place before political democracy could be expected to function. First is personal democracy—a willingness by all to respectfully and thoughtfully listen to other opinions, to exchange viewpoints and engage in dialogue. Second is economic (he called it industrial) democracy—a state of affairs in which grinding poverty was eliminated and basic needs were met. Third is educational democracy—in which illiteracy were eliminated and all had complete and free access to developing their talents and pursuing their interests.[12] Absent these three preconditions, political democracy is apt to become the plaything of powerful pressure groups and, indeed, that is what has happened.

In the United States today, public dialogue about issues that affect the lives of the electorate is regrettably sparse. We tend to live in our enclaves and communicate with our social and educational peers. In part this is due to the lack of the second precondition Dewey mentions: the economic and hence the social situation in the United States is producing a steady growth in extreme poverty as well as in extreme wealth. Furthermore, primary and secondary education, because of the way in which schools are funded tends, to be bad precisely where it should be excellent and access to college and University is more and more restricted to those with higher incomes. This creates a situation in which true political democracy cannot flourish. Indeed, it is one in which a viable political democracy predictably will die.

It is difficult to create a health care system in the context of such a situation. Poverty and lack of education are both directly linked to disease—the lower the income and the lower the level of education the higher the incidence of almost every disease studied. Therefore, we have a task before us: while emphasizing the importance of craftsmanship of any health care system, we must be aware of the social context in which such crafting takes place. If done right one can hope to not only craft a flexible, changeable and equitable health care system but in the process of doing so benefit the entire social system. Taking care that the process is interactive and democratic and not simply a "top-down" strategy can set an example for other social policies to follow.

We who teach health care ethics have an obligation to lead the way. Teaching health care ethics is a social task. Important as the problems at the bedside and the options available to us in dealing with them are, they are inevitably shaped and constrained by the system in which they take place. Again, for too long we who teach health care ethics have been chiefly concerned with what I have called "rich-man's" ethics—the ethical problems those of us well off and able to have ready access to the health care system have had the luxury of being able to worry about. At best we have paid a few moments of lip service to the millions whose problem is not when to stop treatment or whether to get in vitro fertilization but where to get a meal, shelter and care for their hypertension. Worrying about the ownership of a dead man's sperm—an activity that consumed hours of time for persons of considerable talent in a health care ethics discussion group to which I belong—is as "safe" as it is useless. It is an interesting parlor game and one that will not get one into difficulties with the establishment. Keeping oneself safe, sitting in one's ivory tower and studiously ignoring what is going on in the real world is precisely what academics did in Nazi Germany. With that they not only were neutral to any solution, they very much became part of the problem. One would hope that we in today's world and in nations which pretend to be democracies can do better than that.

ENDNOTES AND REFERENCES

1. EH Loewy, *Textbook of Health Care Ethics* (NY: Plenum Publishers), 1996.
2. There are many arguments which have been made for a "right" (or justified claim) for health care. One of the most carefully crafted is N. Daniels, *Just Health Care* (NY: Cambridge University Press), 1985. Daniels argues that health care is needed to achieve fair access to the life plans which any reasonable member of a community might wish to pursue and that to make such plans possible is one of the obligations communities have.
3. In an overlooked but highly significant book Margolit argues that the decent society is the society which shames or embarrasses people as little as possible. See: A. Margolit, *The Decent Society*, trans. by Naomi Goldstein (Cambridge, MA: Harvard University Press), 1996.
4. EH Loewy, *Freedom and Community: The Ethics of Interdependence* (Albany, NY: State University of New York Press), 1995.
5. EH Loewy, "Of Markets, Technology, Patients and Profits, *Health Care Analysis* 1994; 2 (2): 101-110.
6. This brief comparison is abstracted from a recent edition of *Health Care Analysis* in which various systems, their structure and consequent ethical problems are discussed. See: Special Issue of *Health Care Analysis* (ed. by EH Loewy), 1999; 7(4): 309-411.
7. B. Barry, "The Welfare State *v.* Relief of Poverty," *Ethics* 1990; 199: 503-529.
8. For a discussion of this issue, see EH Loewy, "What would a Socialist Health Care System look like? A Sketch," *Health Care Analysis* 1997; 5: 195-204 and EH Loewy, "Justice and Health Care Systems: What would an Ideal Health Care System look like?" *Health Care Analysis*, 1998; 6: 185-192.

9. The "veil of ignorance" introduced by J. Rawls can, despite its drawbacks (which take us too far afield to be discussed here), serve as a heuristic device in making important social choices. (See, J. Rawls, *A Theory of Justice* (Cambridge, MA: Harvard University Press), 1971. Imagine yourself behind a Rawlsian veil of ignorance which would not have you know who you are, how old, well or unwell you might be, what race, or what gender or income group you belong to. You are told only that you will have to choose among three different social goods, only two of which would be guaranteed to you—the third would be up to luck and your own devices. These three would be a) all biological necessities would be guaranteed; b) a full education to develop your interests and talents would be yours for the asking; and c) health care would be fully supplied in case of illness. It is likely that most prudent choosers would choose to forego guaranteed health care in favour of the other two.
10. EH Loewy, *Moral Strangers, Moral Acquaintance, Moral friends: Connectedness and Its Conditions*. (NY: State University of New York Press), 1996.
11. EH Loewy, "Curiosity, Imagination, Compassion, Science and Ethics: Do Curiosity and Imagination serve a Central Function?" *Health Care Analysis* 1998; 6: 286-294.
12. J. Dewey, "Creative Democracy: The Work before Us," *John Dewey: The Later Works 1939-1941*, ed. by JA Boydston (Carbondale, IL: Southern Illinois University Press), 1991.

Chapter 2

Facing Finitude in Health

How the American Aversion to Rationing Thwarts Health Policy

Larry R. Churchill
Professor of Social Medicine and Co-Director, Center for Health Ethics and Policy
University of North Carolina, Chapel Hill
Chapel Hill, North Carolina 27599
e-mail: churchil@med.unc.edu

Key words: health care policy, health care rationing, social justice, fairness, managed care, market forces, efficiency, technology, social determinants of health

1. INTRODUCTION

Few topics in health care have evoked as much confusion or been subjected to as much political demagoguery as rationing. In his new book *Pricing Life: Why It's Time for Health Care Rationing*, Peter Ubel recalls the scene from the Mel Brooks' movie "The History of the World" in which Moses is coming down from the mountain with three heavy stone tablets. Moses begins, "People of Israel, I have here God's fifteen..."—then he fumbles the load, drops a tablet which smashes to bits—"...uh, make that ten commandments." Ubel says it is likely that one of the commandments on the smashed tablets was "Doctors shall not ration" (Ubel, 2000, p. 99).

It isn't only doctors who are leery of rationing. Patients consider themselves wronged and sometimes harmed by rationing. Recently the U.S. Supreme Court heard a case brought by attorneys for Illinois resident Cynthia Herdrich. Herdrich went to her HMO for a pain in her side, but a diagnostic test that might have detected her problem was delayed. Subsequently, Herdrich's appendix burst, requiring emergency surgery. Herdrich and her attor-

Changing Health Care Systems from Ethical, Economic, and Cross Cultural Perspectives, edited by Loewy and Loewy. Kluwer Academic/Plenum Publishers, New York, 2001.

neys claimed that the delay in properly diagnosing her problem was due to the cost-cutting practices of her HMO, and that its zeal for profit-driven efficiency made her physicians negligent in her care. This case is just the most recent in a long line of horror stories from angry patients throughout the country who believe their care is being rationed in order to increase corporate profits, or to fill the pockets of physicians who are paid bonuses to keep costs down.

The Supreme Court unanimously ruled that Herdrich could not use existing federal law to sue an HMO and moreover, in the words of Justice Souter, "inducement to ration care goes to the very point of any HMO scheme...and rationing necessarily raises some risks while reducing others." (*Pegram v. Herdrich*, 2000). One wonders how Souter would have felt had it been his appendix that had ruptured but, of course, the Court was presumably making a judgment based on law, rather than empathy for the aggrieved patient. The Court also suggested that any remedy for the rationing practices of managed care organizations (MCOs) would have to come from Congress. At the time of this writing, any Patient Bill of Rights that includes the right to sue an MCO looks like a long shot at best.

But whether or not patients ever win the right to sue their MCOs for rationing is only a small part of a much larger issue. Indeed, winning the right to sue could be a step backward in the public's awareness since such a right might imply—contrary to Souter's accurate assessment—that any rationing of health care is unwarranted. Such an assumption would blind the general public to the rationing that routinely occurs at a variety of levels in a market-driven health care system. The U.S. has a long and well-practiced habit of suppressing not only the price-rationing of marketed medical services, but the inevitability of some limits—and hence some mode of rationing—in every health care system. So cases like that of Herdrich seem outrageous, at least in part, because the American public has for so long believed that rationing is an avoidable flaw. But of course rationing does occur, and must occur. It is simply not acknowledged as such. And as long as it is unacknowledged, it remains immune from ethical examination until a crisis such as Herdrich's occurs.

In what follows I want to develop a brief history of three phases of American attitudes about limitations to and rationing of health services. This will necessarily be a general overview, intended to capture prevailing sentiments and assumptions, to which there are many exceptions. Still, I believe that this short review can assist us in seeing Justice Souter's remarks in social context and as exemplary of a growing if episodic awareness that health services are rationed, at least under managed care. Whether there is a broad general awareness that rationing is necessarily a part of any health care system, however, is a different matter.

2. LIMITS AND RATIONING: THREE PHASES OF AWARENESS

2.1 Denial

During the 1980s, and also prior to that, we were as a nation largely in denial about rationing. The fear was that escalating costs might eventually bring us to rationing. If medical costs continue to be double the inflation factor, we might end up like Canada and Europe, and actually deny people needed services, or at best have delays and waiting lists for non-emergency procedures. If we could control costs, so the thinking went, we could avoid the moral tragedies of rationing that besiege the so-called "socialist" and second-rate health systems.

In 1984 Henry Aaron and William Schwartz published a book entitled *The Painful Prescription* which typified the thinking of this era (Aaron and Schwartz, 1984). In this volume they described the rationing of hemodialysis in Britain's National Health Service. They were very critical of the age-based allocation scheme that allowed older Britons, usually those in rural areas over 55, to die from renal failure rather than be referred to dialysis units at large urban medical centers. This was not official policy; this was simply accepted practice among British general practitioners in order more efficiently to utilize the scarce resource of hemodialysis. Most observers now think that the U.K. was seriously under-resourced for hemodialysis during this period, and practices of non-referral are now greatly reduced. Aaron and Schwartz cited age-based dialysis rationing in the U.K. as an example of practices that have yet to occur in the U.S., and they concluded that if and when it does come it will be a "painful prescription."

My point here is that Aaron and Schwartz wrote in seeming oblivion of the 37 million Americans who were at that time uninsured and underserved, and who subsequently relied on charity—the group Uwe Reinhardt calls "health care beggars." During this period the term "rationing" meant denying services to those who could otherwise pay for them and would otherwise have them. It did not refer to the effects of market forces in pricing people out of care, or the overall consequences of market distribution when the uninsured go begging, receive fewer services than the uninsured and die prematurely. During the 1980s Americans were still in their salad days when thinking about health care and its limits. Our judgments were especially green about the power of new technologies to create medical need, about the size of the American appetite for better services and for higher levels of wellness and greater longevity, and about how the high degrees of certainty about health status sought by the rich would affect the availability of services for the poor. We were also naive about the ability of the pharmaceutical and medical device industry to shape our appetites and the eagerness with which physicians would utilize these drugs and devices. We thought 10% of the GDP for health was an astonishingly high number, not realizing how quickly and easily we would reach the 14-15% range.

This general naivete about health policy was made a more potent force by being coupled with a heady optimism about our ability to avoid rationing through a variety of means. For example, there was a lot of talk about becoming more efficient by cutting the fat out of medical services. On the list to be trimmed were things such as duplicative diagnostic testing—the excess testing spurred by fear of litigation—and the substantial excess capacity of hospitals. We have since learned that one person's fat is another person's bone. The case of Cynthia Herdrich is only one example in a vast litany of complaints about the efficiency measures of managed care, which both patients and physicians fear have lowered quality and clearly have demoralized health professionals. We have learned the hard way that efficiency is merely a tool, and that efficiency measures may be either good or bad, depending upon the goals they seek to achieve and the means they use to achieve these goals. No efficiency measure can be assessed without first being clear about what goal is being pursued: Efficiency in the service of what? Better care? If so, for whom? And to what end? Care for more persons? Larger profits? Market-driven managed care efficiency looks and feels like rationing to many patients and care-givers, and is at odds with some of the less easily measured reasons that patients seek care and that physicians value their profession. Because efficiency is a means and not an end, it is possible to be efficient in the wrong places, and in pursuit of the wrong goals–as I believe is now often the case. But this more sober and probing understanding of efficiency was not a part of the social conversation in the 1980s.

Another favored prophylactic against the need to ration in the 1980s was the belief that advanced technologies just might bail us out of the cost problem. The rhetoric of Lewis Thomas and his juxtaposition of "halfway technologies" versus "genuinely decisive technologies" was in the air (Thomas, 1973, pp. 33-36). Halfway technologies—such as iron lungs—were not curative. They only slowed the downward course of an illness and were typically very expensive. These partial fixes, Thomas asserted, are primitive and rely on inadequate understanding of the underlying disease mechanisms. The genuinely decisive or true technologies—like the Polio vaccine—are curative, but also cheap and relatively easy to deliver. The hope was that as halfway technologies are replaced by true technologies, costs will go down and the need for rationing could be fended off.

I think we now understand that these great leaps of medical progress are rare. More often than not technologies, even very effective ones, create new problems in their wake. Like a mountain climber who has reached the summit only to gain a vista of new peaks, our great successes in medicine often postpone but do not eliminate the need to make rationing policies. For example, many of the newest drug therapies enabling the elderly to live longer and in many cases more robust lives also add to the burden of paying for services for the elderly, spawning initiatives such as the recent proposals of the Clinton administration for prescription drug coverage under Medicare. In addition, highly desirable but very expensive technological improvements in diagnosis are often available long before therapies are devised, as we are now seeing in the sequellae of the Human Genome Project. So Thomas's thesis seems not to apply to many diagnostic technologies, and only rarely to therapeutic ones.

Notice that both the efficiency and the true technology defenses posed during this period as a prophylactic against rationing were solutions of ingenuity and progress. This signals that the problem of rationing was seen not only as a problem lying somewhere in the future, but also a problem that would be amenable to what Garrett Hardin called a "technical solution" (Hardin, 1968, pp. 1243-1248.). Hardin used this phrase to describe current approaches to the population problem, but it can also be applied to the allocation of medical services. If a problem is defined as "technical," then the right expertise is all that is needed. In other words, during the 1980s rationing was not thought to be a problem that required a fundamental rethinking of values, or a reconsideration of the role and place of health services in society, or any deep probing how the American image of the good life is shaped by utopian expectations of medical care.

I am not, to be sure, claiming that it was somehow a faulty 1980s definition of the term "rationing" that kept us in denial about what was happened all around us. Rather this definition fit a larger cultural disinclination to look at issues of justice in health care. Although society seemed to respond to medical rescue situations, or be horrified by cases such as Ms. Herdrich's, we were loath to examine the systematic cruelty of the patchwork system that had developed. We were too busy identifying the flaws in the Canadian and British systems to see the havoc being created by our acquiescence to the tacit rationing of the U.S. market.

And this brings me to one final point I want to make about this age of denial. The ingenuity of rationing health care by market forces—by price, or its surrogate, insurability—is that no one is to blame for the bad outcomes. Are 37 million (the 1987 estimate) left uninsured and underserved? Do these people suffer greater morbidity and die prematurely? Well, that's too bad, but we didn't decide to exclude them! It's just an unfortunate side effect of the way things work. The genius of using the market to allocate health care is that nobody is in charge of the system, and therefore nobody is to blame for the injustice of the outcomes (Churchill, 1987, pp. 14-15). This displacement of responsibility for the consequences of the system is one of the factors that have allowed our unjust health policies to endure and resist criticism for so long. Letting market forces govern the system reinforces our denial about rationing by uncoupling it from any notion of moral agency.

2.2 Awakening and Anxiety

If the 1980s were a period of denial, the 1990s began as a period of awakening and anxiety. The economy was sputtering. Employers were increasingly concerned about escalating costs; employees were worried about upward mobility and career enhancement moves because of pre-existing conditions clauses in most health insurance policies. The result was "job lock" and career stagnation. In spite of Diagnosis Related Group payment schemes (adopted for Medicare in 1983), inflation for government spending for health care was a major drag on the economy. But the new wrinkle in awareness during this period was that concern for costs was paired with concern for access. A U.S. Senate race in Pennsylvania was decided in part by the differences between the candidates on health policy, and there was growing public awareness that price-rationing was an accurate descriptor of the American system, and that it was unfair. But the awakening was not complete, for it seemed not to extend to the deeper perception that any and all health care allocation schemes will require some rationing because medical needs always outrun available resources.

Yet even this period of limited consciousness and realism didn't last very long as the window of opportunity for reform was soon slammed shut. While many of us became far more aware of just how our patchwork system functions, and more acutely aware of our own vulnerability in a system that depends on employer-based access, the idea that all health care systems have limits, *necessarily* have limits, was a message that was neither sent nor received. The awakening was only partial, for it was confined to a more acute awareness of vulnerability; it did not encompass facing the inherent finitude of health services that any contemporary society faces. Our national leaders did not lead us to achieve this more complete awakening. President Clinton, for example, although very eloquent on the need to cover everyone, said little or nothing about the fact that fairness requires tough allocation choices. He made no mention of the fact that every modern society has to ration—by whatever term it may be called—and that it is impossible to cover all the health care needs of all citizens. In fact when a group of bioethicists were assembled in the early 1990s to help draft the values statement that would accompany the Clinton health plan, they were specifically instructed to draft principles that did not include the word "rationing"!

Contrast this with the approach of John Kitzhaber, then President of the Oregon Senate and a chief architect of the Oregon Health Plan. Kitzhaber was very clear about the need to ration services in order to avoid rationing people (Kitzhaber, 1990, pp. 2-5). Medicaid programs in most states deal with medical cost overruns and dwindling budgets by simply changing the eligibility requirements and making fewer poor people eligible for the full package of benefits. It's all or none; one is either in the system and eligible for all services, or out of the system and eligible for none of them. The Oregon strategy advocated rationalizing benefits based on their effectiveness, and providing this more modest but effective package of services to larger numbers of people.

There is of course much more to be said about the Oregon Health Plan, and parts of it are quite controversial. I cite Kitzhaber and the Oregon Plan at this juncture to emphasize two things I think Oregon did right. First, unlike the Clinton administration, they couched their reform in a realistic assessment of limits on resources and confronted this candidly. Secondly, because the Oregon legislature had to decide which services on the ranked list to fund, the process resulted in making the rationing of health services an open process with public accountability, rather than something hidden in the mechanisms of fluctuating eligibility rules, or market forces.

What can we learn about rationing from this contrast between the Clinton initiative and the Oregon initiative? That an honest acknowledgment of limits, framed within a system of public deliberation and accountability, may in some instances be politically more viable. Several researchers have observed that what the Oregon Plan achieved was not a change in health care delivery, but a change in the politics of health policy (Jacobs, Marmer and Oberlander, 1997). The fact that Clinton's reform proposal shied away from any real considerations of limits to health services and made any talk of rationing off limits left it open to the sort of demagoguery characterized by the insurance industry's famous "Harry and Louise" ads. In these ads the couple are seated around the kitchen table with bundles of papers in front of them (the Clintons' plan was hundreds of pages long). Louise says to Harry, something like "This big government program will take away our choices for doctors and treatments." In effect, the Clinton reform will ration our care. Ironically, more people have fewer choices now under market-driven managed care than would have been the case under Clinton's Health Security Act (Starr, 1994, pp. 70-77; Kuttner, 1999a, 1999b). My point is that the utopian rhetoric of the Clinton reform plan left it open to a Harry-and-Louise-type attack, which implied that rationing was the hidden agenda of the federal government. The strategy of accusing others of sins one commits daily is an old one, and the irony of it has been largely lost on the general public until recently. Rationing has become increasingly visible with managed care, especially the for-profit version, and has found an unequivocal voice in Justice Souter.

2.3 Amnesia

The third period, 1995-1999, has been a period of amnesia. In the vacuum created by the lack of federal action on cost control, the private sector stepped in and has been applying its own measure of efficiency, with mixed results. While some efficiencies have been achieved through managed care, the more accurate picture is not one of cost savings, but cost shifting—shifting costs from MCOs to patients and providers. Yet the chief thing to notice about these last five years is that concern for access has diminished and any ambition for universal coverage has completely dropped from sight. The early 1990s were filled with conferences, workshops and symposia on universal access. I am unaware of any between September of 1994 and September of 1999. The year 2000 has brought a renewed concern for access for the uninsured, but the anxiety of the working middle class so palpable in 1992 no longer animates health policy. As a result, reform proposals that would cover everyone are still off the screen.

The last five years have seen rationing with a vengeance—market rationing and managerial rationing, with little concern for issues of justice or fairness (Churchill, 1999). Congress has largely been in a reactive mode, trying to make small adjustments to mitigate some of the worst effects of market forces. The Kassebaum-Kennedy Act (HIPAA) of 1997, for example, provided for greater portability following job loss or change, but does nothing to restrict the inevitable escalation in premiums for those who are between jobs or changing insurers. In spite of the Children's Health Insurance Program (CHIP), and the efforts in many states to include more persons under their Medicaid programs, the number of uninsured nationally is now at 43-44 million and is increasing at a clip of roughly 1 million per year. Predictions of the number of uninsured are as high as 60 million by 2005, depending upon whether the economy weakens or remains robust. Moreover, the uninsured are likely to receive even fewer services than they did in the past, since cross-subsidization is declining, especially in areas with high managed care penetration (Cunningham, Grossman, St. Peter and Lesser, 1999, pp. 1087-1092). The awareness and anxiety that marked the early 1990s has all but disappeared with the bull economy. Perhaps it will take another recession to awake us to a more realistic assessment of our situation. Then the public might be more open not only to their own vulnerability, but also to the inevitability of limits and the necessity for rationing. Such awareness seems rare in the present climate, and candidates for neither major party are likely to embrace the sort of ethical-political right to health care that characterized the 1992 presidential election.

This twenty-year portrait of the progress and regress of American awareness is very limited, and leaves out many subtle differences among divergent segments of the population, as well as among those who shape U.S. health policy. Still, I think the general portrait is accurate and reveals a deep cultural legacy of denying limits and avoiding issues of rationing. The frontier mentality of the dominant culture that keeps us moving and innovating—often a great asset—also keeps us in denial about non-technical solution problems. Why adjust our attitudes if we think we can fix the problem without the hard work of reflection and reevaluation. Here, at least, our scientific and technological prowess has worked against us. Denial of limits is—culturally speaking—the default position and it has limited our view of the possibilities and thwarted our efforts at reform. Regrettably, as we begin the 21st Century this denial still seems quite robust. William Haseltine, CEO of Human Genetic Sciences, recently expressed it vividly. "Death," he said, "is just a series of preventable diseases." (*New York Times*, October 27, 1999). In a similar way, we have tended to think that rationing is just a series of avoidable shortages. If leaders of corporate science pander to our denial about death, it is small wonder that we are can't come to grips with limits to health care.

3. THE BENEFITS OF FACING OUR FINITUDE IN HEALTH CARE

It might seem that limitations and the need to ration is, in the words of Aaron and Schwartz, a "painful prescription"—a bitter pill that must be swallowed. A world in which there were plenty of medical and health resources to meet all needs is the one we seem to wish for, but it is not at all clear that this kind of world would be best. I will argue in what follows that such a world in fact would not be best and that dealing with finite resources is not only our predicament, but has substantial benefits. I have three main points.

3.1 Safer Medical Practices

Working within acknowledged limits helps to focus attention on what works and what doesn't. It is no secret that when fee-for-service medical practice was combined with indemnity insurance financing there was a tremendous incentive for over-treatment for both patients and physicians. Patients were frequently eager for more treatment because it was assumed that more care was almost always better care. For patients with employer-sponsored insurance the feeling was typically that they had *already* paid for whatever care they might need. So long as deductibles and coinsurance are low, then, there was a sense of entitlement to whatever they might need or want. For their part, physicians had a powerful financial incentive for doing a multitude of invasive diagnostic and therapeutic procedures, and the more invasive and technically sophisticated the intervention, the larger the reward. This mutually reinforcing investment in more services was a recipe for run-away costs and, more importantly for my purposes, a set-up for iatrogenic disease and injury.

By contrast, working within limits should motivate both physicians and patients to carefully evaluate services, and by so doing lessen the risk of iatrogenic illness and injury. In the current investor-owned managed care era, of course, the risk is not so much over-treatment as under-treatment, receiving too few services, as the Herdrich case dramatically illustrates. The goal should be to devise financial incentives for providers and guidelines for MCOs such that they are motivated to provide all appropriate services, but only the appropriate services. This will require careful planning and regulatory oversight. And such oversight should bring a related benefit, viz., standardizing some of the variation in medical practices. If recognition of limits is accompanied by careful peer review, the overall standard of practice should rise. Of course, whether limits and the rationing that goes with it leads to this benefit depends on who devises the standards of practice, and with what purpose in mind. But if acknowledged limits are joined with greater attention to professional standards and clear public accountability, limits will not hinder but help. Medical practices will become safer for patients than any mode of medical practice that assumes resources are unlimited, or that hides rationing under entrepreneurial agendas. This greater safety is something we have yet to achieve in the U.S. It is achievable only when limits are recognized, and when the modes of dealing with limits, viz., rationing, are subject to professional scrutiny in a system of larger public accountability.

3.2 Investing in Social Determinants of Health

Acknowledged limits in health care can also be a good thing if it allows for medical cost reductions and permits expenditures on things more important to health than medical services. In the popular mind health and the provision of direct medical services are all but synonymous. Yet the research on determinants of health over the past decades clearly shows that the chief factors are environmental and social. Clean air and water, sound nutrition and safe housing, education (especially literacy), meaningful community affiliations and social bonds, social status, and adequate self-esteem are actually bigger factors in health than whether a person has secure access to a physician. This is not to say that medicine is unimportant. It is to say that if we want to improve health status in the U.S., we would be wise to invest relatively more in the social determinants and relatively less in direct medical services.

Indeed, it seems that the more we can reduce social inequality generally, the more we can improve the health of the less well off. Social epidemiologists have known this for decades. Medical ethicists are now beginning to get the message and factor it into their thinking about justice issues in health policy. A recent essay by Daniels, Kennedy and Kawachi entitled "Why Justice is Good for Our Health: The Social Determinants of Health Inequalities," provides an excellent and provocative summary (Daniels, Kennedy and Kawachi, 1999, pp. 215-251). International comparisons of overall health status over the past decades indicate the importance of what is called the "socioeconomic gradient." The greater the degree of socioeconomic inequality in a society, the steeper the gradient of health inequality. Simply put, every step a person can make up the socioeconomic ladder is associated with improved health outcomes. Health differences are, then, not so much a product of absolute deprivation but of the relative deprivation within a country. Put differently, health inequalities are not explained by differences in access to health care. This explains why the upper classes in Britain have substantially better health outcomes than the unskilled labor classes, even though the latter enjoy the benefits of the National Health Service, as well as adequate housing and transportation. In the U.S., the states that have the widest income differentials also show the slowest rates of improvement in life expectancy. Daniels, Kennedy and Kawachi conclude:

> Much of the contemporary discussion about increasing access to medical care misses the point. An intersectoral reform will recognize the primacy of social conditions, such as access to basic education, levels of material deprivation, a healthy workplace environment, and equality of political participation in determining the health achievements of societies.

Given this understanding of what factors are most important for health status, limits on medical services would be a highly desirable thing, *if* the savings from more parsimonious use of medical resources are expended on the social determinants of health.

3.3 Limits as a Condition for Human Wisdom

Finally, the human condition of finitude, of living within limits, should not only be acknowledged but embraced, because learning to live within limits makes for a better life. Indeed, such awareness is the precondition for an authentically human life. Our lives are bounded and circumscribed temporally, physically, and in a variety of other ways. Learning not just to tolerate these limits but to embrace them is part of growing up, both for each of us individually, and as a society.

This is an unseasonable thesis, I realize. The American tradition of westward expansion, of freedom from limits, of improvement and progress, predisposes most U.S. citizens to think that they have failed if they do not have more money and possessions this year than the last. The idea that boundaries are an unpleasant inevitability is one thing, but to embrace them as a positive good seems un-American. Politically, limitations are associated with imposed restraints, with tyranny. So the idea that we should want the medical resources that sustain life and health to be limited may seem like settling for a biological tyranny when we can and should be striving to break free from it. As Rudolf Klein puts it, American society seems to believe in the "perfectibility of man" (Klein, 1984), not just politically but medically, and in both areas the perfection for which we strive seems predicated on breaking down barriers and overcoming restraints. It is no surprise that rationing health resources is an unsavory notion in the company of such idealism.

Yet older and more mature traditions of thinking have recognized the virtues of finitude. Jews, Christian, Stoics and Moslems have all recognized—in different ways—limits to life and health as a blessing. Understanding that one's life, health, and all human resources are bounded is perceived in these traditions as the beginning of wisdom. Knowledge of how to live well emerges from understanding that our days are numbered. Efforts toward ever increasing material possessions, personal or professional promotions, or endless health improvements and life extensions are instances of not knowing our true needs (Kass, 1983; Callahan, 1998).

If there is any truth in these ancient traditions, we should be resistant to any notion that limits are not needed, or not a part of any well-run health care system. The great sin of managed care is not that it has treated health resources as limited, but that it has treated them as a limited market commodity rather than a limited public good. The sin of managed care has not been rationing, but rationing in the wrong way, and for the wrong reasons; in a word, rationing unjustly. Health reforms that are critical of entrepreneurial aspects of delivering medical services must be careful not to be nostalgic about the past, or reinvigorate utopian expectations about the possibilities for a rationing-free system, if only we could get the incentives adjusted properly. No such system can exist, and even if it were possible it would not be preferable. We should hope never to be without limits. Any just or fair health policy will have an important place for limits and will develop from this sense of finitude a fair way to ration. Rather than eschew rationing, let us hope that the next national debate on health care has a central place it.

REFERENCES

Aaron, Henry and Schwartz, William. (1984). *The Painful Prescription: Rationing Hospital Care.* Washington, DC: The Brookings Institute.

Callahan, Daniel. (1998). *False Hopes.* New York: Simon and Schuster.

Churchill, Larry R. (1987). *Rationing Health Care in America.* Notre Dame, Ind.: University of Notre Dame Press.

Churchill, Larry R. (1999). "The United States Health Care System Under Managed Care: How the Commodification of Health Care Distorts Ethics and Threatens Equity." *Health Care Analysis* 7.

Cunningham, PJ and Grossman, JM, St Peter, RF, and Lesser, CS. (1999). "Managed Care and Physicians' Provision of Charity care." *Journal of the American Medical Association* 281.

Daniels, Norman, Kennedy, Bruce, and Kawachi, Ichiro. (1999). "Why Justice is Good for Our Health: The Social Determinants of Health Inequities." *Daedalus*, 128 (4).

Hardin, Garrett. (1968). "The Tragedy of the Commons." *Science* 162. (859).

Jacobs, Lawrence, Marmor, Theodore, and Oberlander, Jonathan. (1998). "The Political Paradox of Rationing; The Case of the Oregon Health Plan." John F. Kennedy School of Government, Occasional paper 5-98.

Kass, Leon. (1983). "The Case for Mortality." *The American Scholar* 53 (2).

Kitzhaber, John. (1990). "The Oregon Basic Health Services Act." Oregon State Senate.

Klein, Rudolf. (1984). "Rationing Health Care." *British Medical Journal* 289.

Kuttner, Robert. (1999a). "The American Health Care System: Employed-sponsored Health Insurance." *New England Journal of Medicine* 340.

Kuttner, Robert. (1999b). "The American Health Care System: Health Insurance Coverage." *New England Journal of Medicine 340.*

Pregrcm v. Herdrich. (2000). U. S. Supreme Court. No. 98-1949, June 12, 2000.

Starr, Paul. (1994). *The Logic of Health Care Reform,* revised and expanded edition. New York: Penguin Books.

Thomas, Lewis. (1974) "The Technology of Medicine." *The Lives of A Cell.* New York: Viking Press.

Ubel, Peter. (2000). *Pricing Life: Why It's Time for Health Care Rationing.* Cambridge.

Chapter 3

Health Care as a Right

A Brief Look at the Canadian Health Care System

Eike-Henner W. Kluge
Professor and Chair, Department of Philosophy
University of Victoria, British Columbia
Victoria, British Columbia
V8W 3P4, Canada
email: ekluge@uvic.ca

Key words: Canadian health care, socialized health care, right to health care, public funding, health law

Abstract: The Canadian health care system is a publicly funded system based on the philosophy that health is a right, not a commodity. The system has been able to all provide all qualified Canadian residents with universal access to all medically necessary services. Its establishment was, and continues to be, opposed by organized Canadian medicine. Of late, it has encountered funding problems because of a flagging Canadian economy. Other problems are posed by Canada's constitutional division of powers, its geographic vastness and a move to regionalization of provincial health care administrations. Moreover, aboriginal health lags behind national standards. Still other challenges are posed by recent legal and technological innovations. Nevertheless, despite highly publicized shortfalls in individual cases, the system functions well and is likely to meet these challenges.

1. INTRODUCTION

There are two fundamentally distinct views on the nature of health care: One sees health care as a right, the other construes it as a commodity. This difference is not merely a matter of perspective: It has tremendous practical

Changing Health Care Systems from Ethical, Economic, and Cross Cultural Perspectives,
edited by Loewy and Loewy. Kluwer Academic/Plenum Publishers, New York, 2001.

implications. For instance, a rights perspective tends to be associated with a socialized approach to health care. Further, according to this approach, the very structure of the health care delivery system must ultimately be justified in terms of ethical principles and economic measures can only be used as tools to effect a just and equitable distribution within the system. Finally, the function of a health care system that is constructed on a rights basis is not to generate revenue but to provide a socially mandated service.

On the other hand, a commodity perspective fosters a corporate view of health care. On this approach, health care is a commodity like any other that may be bought or sold in the market place. Accordingly, economic considerations determine the nature, range and availability of health care services, and ethical principles enter the decision-framework only as identifying the socially mandated limits within which all economic activities have to be conducted. Moreover—and this constitutes a crucial contrast to the rights-oriented perspective—the primary function of a commodity-oriented health care service approach is to generate revenue. It just so happens that the revenue-generating method that is adopted focuses in the delivery of health services. The fact that providing these services also meets a societal need is a serendipitous happenstance that may befall any economic enterprise. Moreover, while on a rights-based perspective the failure to deliver otherwise appropriate health care services to everyone on an equitable basis can be characterized as a failure of social duty, no such claim can be made from a commodity-based perspective. Here, the absence—or even maldistribution—of a particular service is merely a reflection of economic forces that render the provision of the relevant services unprofitable.

For most of its history Canada, in concert with the US and most other countries, had espoused a commodity perspective. As a result, the delivery of health care services grew up in an economic climate where, as a rule, health care was available to all and only those who could pay. The only exceptions were cases where physicians and hospitals provided their services gratis and as it were *pro bono.*[1] Although it was far from ideal, this system persisted essentially unquestioned until the Great Depression of the 1930's. At that time, the economic plight of the majority of Canadians fostered a fundamental and grass roots re-examination of the role of government in society, and of the implications of the social embedding of individual persons. A social conscience began to emerge that identified the provision of certain services as central to the function of any morally responsible society. Implicated here were old age pension, unemployment insurance—and, of course, health care.

This shift in social perspective was not lost on politicians. Consequently, the politically more left-leaning parties—in particular the Co-operative Commonwealth Federation (CCF) in Saskatchewan—began to integrate the idea of socialized medicine into their political platforms. This ultimately led to the introduction of universal health and hospital insurance schemes in Saskatchewan (1944). The move proved politically so successful that in due course it was followed by all other Canadian provinces.

This left the provision of health care in the hands of provincial governments. In a sense, this was not surprising. The Canadian Constitution provides that with the exception of the Northwest Territories, the Yukon Territories,[2] and First Nations peoples living on reserves, health care services are a matter of provincial jurisdiction. The only powers that the federal government can exercise in this regard center in whatever conditions it can attach to the transfer of funds from the federal coffers to the various provinces. The federal government's involvement in the delivery of health care, therefore, could only be indirect. It exercised these powers through what ultimately came to be known as established programme financing, whereby the federal government tried to steer the development of provincial health care systems in the direction of uniform standards and conditions of eligibility for individual persons.

This involvement, dependent as it was on treaties made with the separate provinces, was a patchwork affair. In 1957, the federal government regularized its involvement in the one aspect of health care delivery with the passage of the Hospital Insurance and Diagnostic Services Act (1957). This authorized the federal government to contribute up to fifty per cent to the provinces' hospital as well as laboratory and diagnostic services programs, the monies being provided according to a set of formulae that took into account the varying needs of the provinces, their different fund-raising abilities, etc. While it was a large step towards a national health care scheme, the Hospital Insurance and Diagnostic Services Act still fell short of bringing about a truly universal and comprehensive approach to health care because it did not extend to medical services. Private insurances and private facilities continued to exist, and there remained noticeable regional differences in the level of health care and qualitative differences for Canadians with greater economic means.

In 1964 the Hall Report [Hall, 1964; Taylor, 1987], which was commissioned by the federal government with an eye to revamping the health care system, recommended the establishment of a system of equal and universal access to health care. It thus demonstrated an evolving vision of federal involvement. Part of this vision had already been realized with the Hospital Insurance and Diagnostic Services Act. The passage of the National Medical Care Insurance Act in 1966,[3] which was in response to the recommendations of the Hall Report, further completed the picture. This new Act essentially rounded out health care coverage for Canadian residents by forcing Canadian provinces to provide their residents with universal and comprehensive medical services if they were to receive their share of federal funding.

The final step in the establishment of a truly national and universal socialized health care system occurred when the Hospital Insurance Act and the National Medical Care Insurance Act were consolidated in 1984 in the Canada Health Act.[4] This Act mandated that in order to be eligible for transfer payments from the federal government, provincial health care insurance plans would have to satisfy the following criteria:

- *Public administration*—which is to say, the provincial insurance plans had to be not-for-profit and subject to a public audit
- *Comprehensiveness*—which meant that the plans had to include all insured services provided by hospitals, medical practitioners and similar services
- *Universality*—i.e., they had to entitle 100% of the qualified residents of a province
- *Portability*—which meant that the residents of the various provinces could not lose health care coverage from their old province-of-residence before a three-month residence period had elapsed, at which time they would be covered by the medical plan of new province or residence[5]
- *Accessibility*—which stipulated uniform terms of access as well as reasonable compensation for health practitioners

Despite intense lobbying by insurance companies and the Canadian Medical Association, the Act was proclaimed in 1985 [Taylor, 1987]. The system has undergone some changes since its inception. Nevertheless, today all Canadian provinces and territories have a universal health care system where physicians and hospitals provide services according to province-wide schedules and where, with exceptions that will be explained in a moment, no health care services have to be paid for by the patients themselves.

Ironically, the development of a socially-oriented health care service had been supported in the late 20's and early 30's by organized Canadian medicine as a way to provide medical practitioners with guaranteed payment at a time when the recovery of fees was notoriously uncertain and the income of physicians was experiencing a steady decline [Naylor, 1986]. Initially, therefore, the self-interests of the profession and the shifting perspective of society converged in fostering a supportive climate for health care reform. However, even this early support was tinged by the physicians' self-perception as small businessmen[6] who had a right to control the conditions of their own practice like other entrepreneurs. Moreover, there was an early and abiding insistence that socially mandated medical insurance schemes should be available only to indigents and persons below a certain income level. This would allow the profession to continue the practice of two-tier billing, with the upper-income earners making up for the reduced fees that would be charged to the insured. The reason organized medicine did not advocate a similar restriction on eligibility for hospital insurance was that medical incomes were not effected by hospital insurance schemes. Finally, right from the very start, organized medicine fiercely defended its prerogative of professional autonomy by maintaining that control of any medical insurance scheme should reside firmly in the hands of the medical profession [Naylor, 1986].

These professional considerations constituted a constant and abiding counterfoil to the direction of the evolving social consciousness. It was centralized under by the leadership of, and coordinated by, the Canadian Medical Association (CMA), a private corporation set up by an Act of Parliament in 1867 to advance the economic interests of its physician members. In keeping with its mandate and reflective of the professional perspective just mentioned, the CMA challenged the Canada Health Act on constitutional grounds in a last-ditch effort to derail the Canada Health Act's implementation. However, it abandoned its suit after it became clear that it would lose the challenge. It has since shifted away from an open and formal opposition to the principle of socially controlled and universally available health care. Instead, it has concentrated its efforts on touting the alleged advantages of its own original conception of a two-tier health care system, where the majority of people would be covered by universal health care insurance while those in the higher income brackets, who could afford to pay privately, would make their own payment arrangements. Payment for the latter would not be tied to the socially set fee schedules. At the same time, such private payers would receive what are claimed to be higher-grade and speedier services. To date, the efforts of organized medicine have met with uniform resistance both from the public as well as from government at all levels [Roos, Fisher, Brazauskas, Sharp and Shapiro, 1992].

The system of universal health care insurance that is currently in place is not entirely universal. Exceptions to the universal coverage have already been mentioned. They include treatments that are non-standard—experimental medical procedures fall into this category—and those that are not considered "medically necessary." At first glance, this might be construed as a way for public funding agencies to limit the nature and amount of medical services in keeping with some preconceived financial ideal. In fact, however, this is not entirely the case. The provincial governments have established medical service commissions that decide what is considered medically necessary. These commissions are made up of representatives of the respective Ministries of Health, the public and organized medicine within the relevant province. They function at arm's length from government itself. By and large, they have been very successful in maintaining their independence. As was said, the only interventions that the commissions will not fund as a matter of course are dental services,[7] purely cosmetic interventions and the like—as well as experimental treatments. It is felt that, in view of the unproven nature of experimental procedures, public funds should not be spent on them. However, if prospective recipients of experimental treatments can successfully make their case to the responsible Minister of Health, funding for such treatment may be forthcoming on an *ad hoc* basis.

This picture is currently undergoing some changes at the level of the provinces, with a move towards regionalization. The provincial Ministries of Health have begun to divide their respective jurisdictions into demographically identified regions, each with their own regional health board. The regions are assigned global budgets and the regional health boards have the power to decide over health care planning and delivery within their respective jurisdictions. The driving assumption that underlies this change is that devolving such decision-making power to the regional level will increase flexibility in meeting differential regional needs, while at the same time retaining the same level of care. However, even under regionalization, the ultimate co-ordinating and financial responsibility remains with the provincial ministries of health.

Alberta has recently passed legislation permitting private clinics to offer services on a competitive basis with publicly funded health care institutions. This has been perceived by some as a crack in the health care delivery structure established by the Canada Health Act. However, the Alberta legislation stipulates that, to be able to provide what are provincially insured services and to be eligible for public funding, such private institutions must underbid public institutions with respect to these services. As to the remainder of the services offered, these would be non-insured services and hence fall outside of purview of the Canada Health Act.

2. SUCCESSES

The Canadian health care system can point to many achievements and successes. Chief of these is that by and large, it has succeeded in doing what it was set up to do: namely, to provide appropriate health care to all who need it on an equitable basis without plunging the recipient or significant others into financial crisis [Anderson, 1997; Gironimi et al., 1996; Nair et al., 1992; OECD, 1994; Wilkins and Brancken, 1989; Roos and Mustard, 1997; Rublee DA, 1989]. There is no parallel to the "cream-skimming" and hence the under-servicing that is experienced with private for-profit health service suppliers, nor is there a financial incentive to under-service those who suffer from severe or protracted illnesses in order to maximize profits.

Since in a socialized system profit is not an issue, whatever burdens exist will be spread over society as a whole. This is especially important with an aging population and in end-of-life situations, where the financial implications might otherwise lead to an abbreviation of services. It also permits the current debate over assisted suicide and euthanasia to be focused solely on the issue of autonomy and quality of life, rather than on the financial fears of next-of-kin and significant others.

Further, there is no disparity in the quality of services that are provided to different socio-economic classes[8] since private health care providers—in so far as they exist—may not charge more for the same insured service than public providers. The only exceptions are services that are not covered by the provincial insurance plans. In these cases, fee schedules are at the discretion of the private insurer. However, since all medically necessary services are covered by the provincial insurance plans, this involves only a small minority of services. Usually add-ons such as private rooms (rather than two- or four-person shared rooms) when this is not medically necessary, top-grade as opposed to standard prosthetic implants, name-brand drugs when generic drugs are just as safe and effective, etc. are involved.

The overall effect of the general policy has been excellent. Not only has it created an equal standard of care for all citizens, it also tends to remove financially-based differences in the quality of professional services that are offered by different institutions. Since all health care institutions can only recover according to the same fee schedule, no institution can offer substantially better salaries in order to attract better-qualified health care professionals. Consequently, the pay-differential between privately ad publicly funded institutions that exists in some other countries does not obtain. The only significant differences between institutionalized health care involve an institution's affiliation with a medical school or its designation by the public funding agency as a primary as opposed to, say, a tertiary health care facility. However, even here, any differences that might impact negatively on the quality of care received tend to be smoothed out by societally funded patient-transfer mechanisms that shunt patients with higher health care needs to the appropriate tertiary institutions. The system of medical transport in Canada is very highly developed.

As to the quantity of services, there is no pre-set limit as to the number of times that anyone may access the health care system, nor is there any limit on the types of services that are available. The only exception to this obtains in the sector of experimental treatments, which will not be available. Initially, there was some fear that the easy availability of services would lead to consumer abuse. However, that does not appear to have materialized.

There are waiting lists for certain procedures and interventions. Consequently, there have been some efforts, spearheaded by organized medicine, to establish private clinics to allow people who are financially well off to circumvent waiting lists for publicly funded services. However, these efforts are few and far between and generally involve services that are not considered medically necessary.[9] Moreover, federal transfer payment guidelines under the Canada Health Act and related legislation contain penalty clauses that focus on differential payment schemes for provincially insured services. They penalize the provinces if they allow private clinics to charge more than the established reimbursement rate that is set by the medical services commissions and paid to not-for-profit institutions. The penalty consists in the withholding of federal transfer funds in the amount that is charged extra by the private service providers. By and large, the effect of this has been that the provinces have adopted a negative stance towards the proliferation of such clinics. Thus, while there are private clinics that offer laser eye surgery, none offer transplants, tonsillectomies, chemotherapy and the like.[10]

Finally, Canada's socialized approach to the delivery of health care has shown itself to be administratively more efficient than private provider-based medicine in other countries and, in particular, the U.S. There is no agreement on the exact amount of this difference, but generally accepted figures range from 2 to 5.5%. With health care budgets running into the billions of dollars on an annual basis, this figure is significant at the level of hands-on care that can be provided. It entails that the Canadian system can use this money to fund treatments where otherwise it would be spent on bureaucracy [Woolhandler and Himmelstein, 1991; Woolhandler and Himmelstein, 1997].

3. ISSUES AND PROBLEMS

Alongside its successes, the Canadian health care system has its share of problems. These can be grouped under three headings: financing, delivery of services and ethico-legal issues.

3.1 Financial Issues

Possibly the most egregious problem lies in the area of finance. Over the last decade or so, the federal government has increasingly pulled back from direct involvement in health care funding. In a series of moves culminating in the *Government Expenditures Restraint Act* (entering into force in 1991), the federal government amended the *Contributions Act*—which controls the transfer of federally collected revenues to the various provinces—by tying these payments to an increase in the gross national product. In times of inflation, this has the effect of successively reducing the amount of money that is actually transferred. This has forced provinces to look elsewhere for the funds necessary to run their health care services. Since provincial funding is dependent on taxes, the ability of the provinces to make up for lost revenue has been severely limited. Provinces whose economies have seen a slow-down in economic growth in recent years—and to some extent this includes most Canadian provinces—have been especially hard hit by this. In this connection, Alberta's recent move to legalize certain private clinics must be seen as an attempt to cut health care costs by introducing competition into the public sector, not as an attempt to undermine the Canada Health Act.

Unfortunately, the federal government has essentially abandoned so-called Established Programme Financing. This means that federal transfer payments are no longer designated to specific programmes such as health care but are lumped together with other financial transfers. This has allowed provincial governments to place these funds into their general revenues, which are then distributed according to perceived financial exigencies and in accordance with a politically motivated process that varies from province to province. The effect has been some erosion in the previous inter-provincial uniformity of services. However, the federal government has recently moved to counterbalance this by increasing the amount of federal contributions. Unfortunately it has done so without reintroducing Established Programme Financing in its previous form. Current federal budgetary surpluses have resulted in increased federal health care spending. This has the potential for reversing the erosion of *de facto* federal regulatory powers with respect to inter-provincial standards of services. Still, the effects of previous federal policies have not been completely ameliorated; in fact, they have been exacerbated by a concerted move on the part of most provinces towards balanced financing. Cutbacks in provincial health sector funding have been implicated.

Further, there has been a move to restructure the transfer payment arrangements on a wholesale basis with a renewed "social contract" arrangement between the federal government and the provinces. However, one of the problems standing in the way of any rearrangement is Quebec's insistence on managing all public expenditures without any interference from the federal government. This demand has been integral to Quebec's political posture for some time. It is tied to the *Parti Québécois's* attempt to take Quebec out of the Canadian federation—which is why health care funding has been an integral part of constitutional debates for several decades. This creates the illusion of uncertainty surrounding the delivery of services, whereas in fact both level and quality have remained uniformly high [Nair, Karim and Nyers, 1992; OECD, 1994].

3.2 Delivery of Services

3.2.1 Waiting Lists

Closely associate with financial constraints is the issue of waiting lists for insured services. Because public resources are limited, there are waiting lists for some services, and in some cases the Canadian waiting lists for these services are longer than in certain other countries [De Coster et al., 1998; Dunn et al., 1997; Naylor et al., 1997; Katz, Mizgala and Welch, 1991]. This has been portrayed in some quarters—especially by the Canadian Medical Association—as resulting in a mass Canadians exodus to the US for medical services, and as indicating the unworkability of a Canadian style single-payer approach.

However, studies have shown that despite waiting lists, the health outcomes for Canadians are no different from countries in which waiting times are shorter—and that in some cases they are better. Moreover, contrary to the dire claims made by organized medicine, it turns out upon investigation that Canadians are not seeking health services from other countries in record numbers [Katz, Verrilli and Barer, 1998]. Finally, although waiting lists have rekindled the drive by professional medical associations for permitting private health care providers to enter the field, there has been no corresponding public support for this. Instead, public pressure has been in the direction of increased public funding [Evans and Roos, 1998]. Consequently it seems that financial exigencies notwithstanding, the Canadian public remains firmly committed to the principle of socialized medicine. Interestingly enough, other health care professions such as nursing have not thrown their support behind organized medicine. Instead, they have supported the public in its demands for increased funding. Finally, alternative health care professionals such as chiropractors, homeopaths, etc., who theoretically might have been expected to be on the side of the CMA, have consistently been on the public side of the debate and instead have argued for inclusion in the socially funded system. To some degree, they have been successful: Chiropractic services are a limited insured item in most provinces.

3.2.2 Professional Dissatisfaction

Nevertheless, it is undeniable that over the past few years, the restructuring of the payment relationship between the provinces and the federal government, and the reduced involvement of the federal government in direct health care funding, have led to a diminished availability of health dollars, both comparatively and in adjusted terms, at the provincial level. This has reduced the maneuverability of the provinces with respect to rising health care expenditures. This, in turn, has fostered professional dissatisfaction on two fronts: once with respect to the level of professional fees and salaries, and once with respect to the ability of the relevant professionals to practice in what they consider to be a professionally responsible fashion.

In the case of nurses, reduced funding has meant small pay increases or no pay increases at all, as well as a reduction in the number of nurses that are hired. Nurses therefore feel that to the insult of effectively reduced pay is added the injury of an increased workload. Consequently some nurses associations have undertaken job action to force their provincial governments to loosen the purse strings. Their efforts have met with some success.

There has also been a reduction in the number of new nurses entering the labor force, as well as some exodus of nurses either out of the profession entirely or to the United State to seek better working conditions. In the absence of a sizeable influx of money to deal with these issues, it can only be expected that these trends will continue. At the same time, it should be noted the provincial governments are apparently beginning to respond to the situation. Some of them have increased their budgetary allowance for nursing care by several tens of millions of dollar. For a comparison to the US, this should be understood with a multiplier of at least ten.

Physicians are dissatisfied for reasons that are partly similar to those of nurses: Their reimbursement schedules have not increased at the rate they would like, and health regions have had to scale back their capital expenditures for equipment, programmes, etc. Consequently physicians feel that their services are no longer valued at an appropriate level. Furthermore, some feel that they are forced to practice in a professionally dubious fashion because they cannot offer their patients all of the services that, in their professional opinion, the patients should receive—or because they cannot offer these services in what they consider to be an appropriately short time-period. However, while the nurses associations perceive the solution to these issues to lie in increased funding for the system as a whole, the professional medical associations claim that nothing short of a fundamental revamping of the socialized approach to the health care system as a whole will work.

In terms of the philosophical dichotomy outlined in the beginning, organized Canadian medicine continues to maintain the contradictory position that health care is both a right and a commodity: a right for those who cannot pay, a commodity for those who can. Organized medicine therefore continues to advocate a two-track approach, according to which a privately funded health care system should be allowed to function in parallel to the current socialized system. Under the leadership of the CMA, it is continuing to lobby for its own historical vision of a two-tier health care system in which individual physicians can retain the ability to serve all who need health care but at the same time reap the profits of higher fees from the financially more able.

It is noteworthy that societal support for the positions of the two professions is by no means uniform. The nurses' position has struck a responsive chord in the public. Nurses are seen as the defenders of patients' interests and as genuinely dedicated to the idea of a socialized health care system with equal access for all. That is why their efforts to secure better working conditions have generally met with solid support.

By contrast, organized medicine's position has found little acceptance. This is probably because its demands are generally seen as self-serving. The public apparently remembers the CMA's opposition to socialized health care and its challenge of the Canada Health Act. Further, the public may well be puzzled by physicians' demands to be seen as business persons in a free enterprise system when provincial governments defray a large proportion of physicians' malpractice insurance costs as part of the negotiated settlements with provincial health care associations, and when the provincial governments provide a significant annual education stipend to each physician who is practicing in their respective jurisdictions. At a deeper level, however, the lack of public support probably lies in a fundamental divergence on the philosophy of health care. As was noted in the beginning, whereas the public has firmly espoused a rights-perspective, organized medicine has retained a commodity-perspective and apparently acknowledges the rights-perspective only insofar as it does not interfere with the advancement of the interests of organized medicine itself. The public also seems to be convinced that while its current socialized health care system may have its problems, these can be fixed by appropriate funding remedies that do not require the sacrifice of the system itself [Evans and Roos, 1998].

3.2.3 Geographically Based Problems

Another pragmatic problem lies in the demographics of health care delivery. Canada's vast size and relatively small population make the delivery of health care in outlying areas very expensive. Efficiency and cost minimization clearly favor a centralization of the delivery systems. However, the belief in social responsibility for equitable access to health care militates against restricting access to only to those who happen to live in large urban centers. Consequently Canada has had to develop a system of advanced transportation modalities that allows anyone in need of medical attention, anywhere within the Dominion, to reach a tertiary care facility in a minimum of time. Further, and for the same reason, Canada is increasingly utilizing the tools of tele-medicine—which is to say computers, electronic data transmission modalities and electronic communication technology—to provide assistance to health care professionals in outlying communities.

Still, both of these present their own problems and challenges. Transportation is no substitute for closeness of location, especially when time is of the essence. Further, the most sophisticated access modalities in the world are helpless in the face of the vagaries of an extreme climate that sometimes makes the use of any mode of transportation impossible.

Further, the integration of long-distance transportation modalities into the health care delivery system presupposes an accompanying array of support modalities such as information transmission, and a standardization of the relevant protocols. This is not always the case. For instance, the provincial health care providers have been reluctant to establish electronically accessible databases that would allow emergency response personnel to access the medical records of patients while in transit. This reluctance is not without financial implications in terms of unwanted and unwarranted emergency interventions. Further, it constitutes a source of ethical conflict for paramedics who, in the absence of appropriate records, must follow resuscitative protocols as a matter of regulation. At the same time, it impedes the ability of emergency response personnel to provide as complete and as efficient a service as would be possible if all relevant patient data were known. However, it should be noted that on a comparative basis, Canada still fares extremely well in relation to most other countries.

As to tele-medicine, while the technology has proven itself in many instances, it has raised questions that center in the privacy and security of medical records. There are no standardized protocols at the present time, and the law provides little guidance since it lags behind current developments in technology. While tremendous efforts are being made to deal with the situation, it will be years before anything like a coherent, usable and systematic system of regulations emerges. In the meantime, the various regional health authorities have to do the best they can.

3.2.4 Aboriginal Health

Another issue of grave concern is aboriginal health. As was noted previously, all Canadians are entitled to equitable access to health care. This includes native populations. In their case, health services are not provided by the provinces but are under the direct control of the federal government. The latter has been diligent in funding health care services for native populations. Nevertheless, the rate at which Canada's aboriginal population suffers from disease and debility is almost three times higher than the non-native population (Ng, 1996). The causes of this are not entirely clear. However, indications are that the social conditions under which many native populations live are a major determinant. Consequently the differences in health status cannot be corrected without dealing with these underlying social issues. Efforts are under way to ameliorate the situation. They include treaty agreements that would give native populations jurisdiction over traditional territories and control over their own communities and resources.

3.2.5 Administratively Based Problems

A new type of issue that is beginning to merge has nothing to do with levels of funding or professional-provider conflicts. It centers in the restructuring of Canadian health care delivery in terms of health regions. Over the past few years, the provinces have re-organized their health care delivery structures in such a way that it is no longer steered centrally and directly by the respective ministries of health. Instead, planning and delivery responsibilities have been devolved onto regional Boards who are responsible for the quotidian delivery of health care in their catchment areas. In some instances, these Boards have adopted a Carver style of administrative accountability [Carver, 1990]. On this model, a Chief Executive Officer who is hired by the Board functions as liaison between the Board and all committees, stakeholders, etc. Since the obligations of the Boards and their CEO's are fundamentally distinct, this sets the stage for serious ethical conflict that is not without its implications for the delivery and availability of health care.

That is to say, the mandate of the various regional Boards is to look after the health care needs of the regions for which they have been appointed by the provincial government and for whom they function. This means that the Boards stand in a fiduciary relationship towards the public in their respective administrative areas. The CEO's, on the other hand, are retained by the Boards and owe a primary duty to these Boards. Consequently they do not stand in a fiduciary relationship towards the public. Instead, they are duty bound to function in a professional capacity *vis-à-vis* the administrative tasks given to them by their Boards. Their professional mandate, therefore, centers not in the right to health care but in the delivery of services in an appropriate and business-like fashion. Consequently, there is a professionally grounded drive for CEO's to run their health regions in a business-like and corporate manner, much on the model of an HMO.

The move towards a Carver oriented administrative structure by many health care regions therefore sets the stage for potentially irresolvable conflicts in the administrative functioning of the health regions themselves. Business-like practices, which necessarily view health care as a commodity, run head-on into health claims that find their basis in the rights perspective governing the operations of the Boards and expectations of the health care consuming public. The conflict is all the more insidious because the CEO's are the conduits through which the Boards interact with their stakeholders. Consequently the Boards are extremely susceptible to the corporate orientation that the CEO's are duty-bound to bring forward as a matter of their professional obligation towards the Boards themselves. This move to a Carver-style administrative structure for health regions has been undertaken in the interest of maximizing the availability of health care through appropriate business practices. Ironically, it has an even greater the potential for undermining the spirit of the Canada Health Act than the attempts of organized medicine.

3.3 Law and Ethics

A still different set of problems derives from the social parameters. That is to say, health care, like any social undertaken, is embedded in a framework of laws and regulations that delimit how it may be delivered. In Canada, as in any common law country, this legal framework is determined by statutes as well as case law. In an ideal society, there would a congruence between the legal framework within which health care is delivered and the ethical parameters and principles that ought to govern the delivery of health care itself. Canadian society is no more and no less ideal than any other. Consequently, there are occasions where the fit between the legal framework within which health care operates, and the ethical requirements inherent in the mandate of health care itself, is problematic.

3.3.1 Conflicts

This absence of congruence takes two forms. In the one kind of case, the law—or at least some part of it—is in direct conflict with itself or is at variance with what society considers ethically appropriate. In the other kind of situation, there is no law at all and the very absence of a law leads to ethical dilemmas.

An example of the first kind of situation is provided by the fact that most Canadians believe that competent persons have the right to execute advance directives detailing their treatment once they are no longer competent to make such decisions in their own person. There even is case law to the same effect (*Malette v. Shulman*, 1990). However, in most provinces statute law and its accompanying regulations have not caught up with this change in ethical perspective. Consequently health care delivery protocols tend to require emergency response personnel to initiate resuscitative treatment even when there is an advance directive to the contrary.[11] This sets up an ethical dilemma for emergency response personnel which, under the circumstances, is not conducive to the delivery of good and efficient health care. Emergency responders are forced to choose between what is ethically appropriate and legitimate (and is recognized as such by case law), and what is required by statute and regulation.

Another example is provided by the prohibition of assisted suicide. Suicide was decriminalized in Canada in 1972. However, assisted suicide remains an offence under s. 241 of the *Criminal Code* and is punishable upon conviction for up to fourteen years of imprisonment (*Criminal Code,* C.c-46 s. 241). This statute stands in flagrant opposition to the ethics of autonomy, and is in direct conflict with the position accepted by the majority of Canadians [Lavery, *et al*., 1997]. It also contradicts the opinion of a significant number of physicians [Verhoef and Kinsella, 1996]. However the Supreme Court, while unanimous in its finding that s. 241 violates the equality and justice section of the *Canadian Charter of Rights and Freedoms*, nevertheless found by a bare majority that the current prohibition is constitutionally validated by s. 1 of the *Charter.* This section allows for a suspension of individual rights if this "can be demonstrably justified in a free and democratic society."[12] It thus introduces an ethical tension into the delivery of health care at the end of life. The situation is anything but ideal.

3.3.2 Absence of Law

The second sort of the lack of congruence between law and ethics lies not so much in an opposition between the two but in a total absence of law in a particular area where, ethically, there should be a legal determination. A good example of this is the issue of abortion. Ethically, there appears to be general agreement among Canadians that after a certain period of gestation, the human fetus attains a morally significant status that should be taken into account when the issue of abortion is broached in the physician-patient interaction.[13] However, since the previous abortion statute of the *Criminal Code* [*Criminal Code of Canada*, s. 251] was struck down as unconstitutional in 1988 (*R. v. Morgentaler* [1988]), there is no abortion law in Canada. Consequently it is entirely legal to perform late second- and early third-trimester abortions. This is a matter of grave concern to many members of the public and the medical profession. This concern appears to be independent of any religious convictions. However, it is unlikely that, in the current political climate, a new abortion law will be introduced any time soon.

The issue of reproductive technologies provides another example. At present, reproductive services are not regulated in Canada except insofar as they fall under the general rubric of the provision of health care. The majority of Canadians nevertheless believes that there *should* be such regulation [Royal Commission on New Reproductive Technologies, 1993]. The federal government has recently moved to fill this gap by introducing criminal legislation to control reproductive technologies on a nation-wide basis (Bill C-47, 1996-97). There are serious flaws with this legislation, not the least of which being that for certain services such as IVF with donor eggs, it would effectively and unalterably discriminate against poorer women. A more general, however, and more fundamental objection that has been raised is that in any case, the criminal law is the wrong tool to use in this instance.

3.3.3 Consent and Children

Another area of ethical concern pertains to children. Canadian society has come to accept that discrimination on the basis of age is ethically unacceptable. This position has found reflection in s. 15 of the Canadian *Charter of Rights and Freedoms* which, *inter alia*, specifically prohibits discrimination on the basis of age. Passage of the *Charter* has required the provinces to amend their consent legislation such that children now may act as decision-makers in their own right, independently of parental veto, as long as the children are deemed competent and the attending health care professional agrees that the choices they make are in their own best interest.[14] Overall, the effects of these changes have been beneficial. Children have been integrated into the decision-making processes surrounding their health [Harrison et al., 1997]. Nevertheless, for some health care providers (especially physicians) it has led to tension between the ethics of autonomy, which has become a cornerstone of Canadian medical practice, and the tradition of paternalism towards children. It has been particularly difficult for physicians who assume that children are incompetent by definition and that parental authority is fundamental under all circumstances. It has also created problems for the mechanics of medical-parental interaction. These are still in the process of being worked out.

4. CONCLUSION

Canadians believe in the principle of equality and justice. That is why Canadians have come to view health care not as a commodity but as a right. This is what underlies Canada's move to a not-for-profit, publicly funded health care system that makes all medically necessary and appropriate health care available to everyone on an equal basis. Because this belief in social equality and justice is an integral part of the worldview of its citizens, Canada's socialized health care system was not imposed from the top down by the federal government. Instead, it was developed from the bottom up, as a result of social pressure. Since it is thus reflective of a deeply ingrained societal attitude rather than a creature of political perspective, it is highly likely that it will survive the challenges that currently face the present system. Indeed, since it is shared by the peoples of all provinces, it would probably survive even a Quebec separation.

Nevertheless, the pressure to effect some sort of change in the current set-up is steadily mounting. In an ideal world, all provinces would have the same population mix and health care needs—and above all the same resources. They could then deliver the same level and degree of services. Unfortunately, this is not the case. There exist great disparities between the provinces in this regard. This makes it very difficult to fulfil the Canadian dream of just social health care unless there is a central coordinating authority regulating and, above all, enforcing authority. To date, that dream has been realized because of the funding role of the federal government and the commitment of provincial and federal governments to the proposition that health care is a right, not a commodity. However, the increasing penetration of commodity-oriented administrative practices and the insistence by organized medicine that the current difficulties would be solved with the introduction of a parallel and private system of health care, exert pressures that are becoming increasingly difficult to ignore. So far the public, which ultimately determines governmental policy, has resisted the siren song of a two-tier system. If public opinion on this issue ever changes, it will spell the end of an historically serendipitous but ethically admirable system of health care.

ENDNOTES

1. Cf. CD Naylor, *Private Practice, Public Payment: Canadian Medicine and the Politics of Health Insurance*, 1911-1966 (McGill-Queens University Press: Kingston and Montreal), 1986.
2. Some of these territories have recently undergone division into distinct judicial entities—e g.. Nunavut—which look after their own health care like the provinces themselves.
3. Revised Statutes of Canada, chap. 64, sect. 4(3).
4. Canada Health Act, S.C. 984, C. 6.
5. The Act also mandated that out-of-country health services would be covered, but only at the respective provincial insurer's rate. This made Canadians attractive health care customers for U.S. health care providers since, a potential fee differential notwithstanding, payment by Canadians was essentially guaranteed—something that was not always the case with U.S. patients.
6. Naylor, 87.
7. With the exception of dental services provided in hospitals and dental interventions for reconstructive purposes.
8. The exception to this is native health. For more on this, see *infra*. However, it should be noted that the native health issue is independent of the issue of private v. public funding for health care since in Canada native health care is publicly funded.
9. The exceptions are private clinics that provide insured services for Workmen's Compensation Board cases. The assumption here is that the sooner a worker can be returned to

work, the less the burdens on society in terms of time-off and sick leave. Consequently the exception is allegedly justified by strictly utilitarian considerations focused on the aggregate social good. Nevertheless, the ethics of this has not gone unopposed and is currently under review.

10. For Workmen's Compensation Board exceptions, see *supra*.
11. For instance, British Columbia has statute law which essentially follows the guidelines set out in *Malette v. Shulman*. That law was passed in 1993 but has not yet received royal assent. Consequently it is not in force, and emergency response protocols follow the attempt-resuscitation-unless-the-patient-is-obviously-dead rule. Most other provinces lack appropriate statutes that follow the lead of *Malette*.
12. *Constitution Act*, R.S.C 1985, Appendix II, No. 44 Part VII Schedule B, Part Of s.1.
13. The figure of 20 weeks is generally accepted as denoting a significant change in status. Thus, the CMA policy on abortion states that abortion is the deliberate termination of pregnancy prior to fetal viability or 20 weeks, whichever comes first (CMA Policy Summary on Induced Abortion, 1176A). For a similar position but couched in general terms, see the concurring minority opinion of Wilson J. in *R. v. Morgentaler* [1988] 1 SCR 30.
14. This has not led to a wholesale rejection of health services by minors, as was initially feared, but mainly to their requesting medically appropriate services such as immunizations, birth control devices, etc. when parental beliefs would otherwise deny them such services.

REFERENCES

1. GF Anderson. (1997). "In search of value: An international comparison of cost, access, and outcomes." *Health Affairs*, 16(6), 163-1711.
2. Bill C-47, "An Act Respecting Human Reproductive Technologies and Commercial Transactions Relating to Human Reproduction." Second Session, Thirty-fifth Parliament, 45-46, Elizabeth II, 1996-97.
3. *Canada Health Act*, S.C. 984, C. 6.
4. John Carver. *Boards that make a Difference: A New Design for Leadership in Non-profit and Public Organizations* (San Francisco: Jossey-Bass Publishers), 1990.
5. *Constitution Act*, R.S.C 1985, Appendix II, No. 44 Part VII Schedule B, Part I s.1.
6. *Criminal Code*, R.S.C. 1985, C.c-46.
7. C. DeCoster, KC Carriere, S. Peterson, R. Walld, L. MacWilliam. (1998). Waiting Times for Surgery in Manitoba (Winnipeg, Manitoba: Manitoba Centre for Health Policy and Evaluation).
8. E. Dunn, C. Black, J. Alonso, JC Norregaard, GF Anderson. (1997). "Patients' Acceptance of Waiting for Cataract Surgery: What makes a Wait too long?" *Soc. Sci. Med.*, 44,1603-1610.
9. R. Evans and NP Roos. (1998). "What's Right about the Canadian Health Care System?" *Toronto Star*, September 21.
10. G. Gironimi, AE Clarke, VH Hamilton, DS Danoff, DA Bloch, JF Fries, Esdaile. (1996). "Why Health Care costs more in the US: Comparing Health Care Expenditures between Systemic Lupus Erythematosus Patients in Stanford and Montreal." *Arthritis Rheum*, Jun.39 (6), 979-87.
11. Mr. Justice Emmett Hall. (1964). *Report of the Royal Commission on Health Care Services* (Ottawa: Queen's Printer.

12. C. Harrison, NP Kenny, M. Sidarous and M. Rowell. (1997). "Involving Children in Medical Decisions," *CMAJ* 156(6), 826-8228.
13. *Health Policy Studies* (1994). 1(5) Table 4.
14. Hospital Insurance and Diagnostic Services Act (1957) Revised Statues of Canada Chapter 28.
15. SJ Katz, HF Mizgala and HG Welch. (1991). "British Columbia Sends Patients to Seattle for Coronary Artery Surgery: Bypassing the Queue in Canada." *JAMA* 266(8), 1108-11.
16. SJ Katz, D. Verrilli, ML Barer. (1998). "Canadians' Use of US Medical Services," *Health Affairs* 17(1), 225-35.
17. J. Lavery, BM Dickens, JM Boyle, PA Singer. (1997). "Euthanasia and Assisted Suicide," *Can Med. Assoc. J* 156, 1405-8.
18. *Malette v. Shulman* (1990) 72 O.R. (2d) 417 (C.A.) 17.
19. C. Nair, R. Karim, C. Nyers. (1992). "Health Care and Health Status: A Canada-United States Statistical Comparison." *Health Rep* 4(2), 175-83; OECD. The Reform of Health Care Systems: A Review of Seventeen OECD Countries.
20. D. Naylor, P. Slaughter, K. Sykora, W. Young. (1997). "Waits and Rates: The 1997 ICES Report on Coronary Surgery Capacity for Ontario. Toronto." The Institute for Clinical and Evaluative Sciences in Ontario.
21. E. Ng. (1996). "Disability among Canada's Aboriginal Peoples in 1991." *Health Rep* 8(1), 25-32; 25-33.
22. OECD. (1994). "The Reform of Health Care Systems: A Review of Seventeen OECD Countries." *Health Policy Studies* 1(5) Table 4.
23. *R. v. Morgentaler* [1988] 1 SCR 30.
24. *Revised Statutes of Canada*, chap. 64, sect. 4(3).
25. *Rodriguez v. British Columbia* (Attorney General), [1993] 3 SCR 519.
26. LL Roos, ES Fisher, R. Brazauskas, S. Sharp and E. Shapiro. (1992). "Health and Surgical Outcomes in Canada and the United States." *Health Affairs* 56-72.
27. NP Roos, CA Mustard. (1997). "Variation in Health and Health Care use by Socioeconomic Status in Winnipeg, Canada: Does the System Work well? Yes and No." *Milbank Quarterly*, 75(1), 89-111.
28. DA Rublee. (1998). "Medical Technology in Canada, Germany, and the United States." *Health Affairs* (Millwood) 8(3), 178-8 1.
29. M. Taylor. (1987). *Health Insurance and Canadian Public Policy: The Seven Decisions that created the Canadian Health Insurance System and their Outcomes.* (Kingston and Montreal, McGill-Queens University Press), second edition.
30. R. Wilkins, D. Adams, A. Brancken. (1989). "Changes in Mortality by Income in Urban Canada from 1971 to 1986." *Health Rep* 1, 137-74.
31. S. Woolhandler, D. Himmelstein. (1997). (Costs of Care and Administration at For-profit and Other Hospitals in the United States." *NEJM* 336, 769-74.
32. S. Woolhandler, D. Himmelstein. (1991). "The Deteriorating Administrative Efficiency of the US Health Care System." *NEJM* 324, 1253-58.

Chapter 4

The Oregon Health Plan Ten Years Later

Ethical Questions and Political Answers

Michael J. Garland
Professor and Vice-Chairman, Department of Public Health and Preventive Medicine
Associate Director of the Center for Ethics in Health Care
Oregon Health Sciences University
Portland, Oregon 97201
e-mail: garlandm@ohsu.edu

Key words: health care policy, health care systems, access to health care, resource allocation, rationing, social solidarity, social justice, fairness, common good

Abstract: Politics and ethics come together in Oregon's experiment with allocating resources for health care. The Oregon Health Plan rests on a commitment of solidarity with the poor. The concepts of the common good, fairness, prudence and wisdom reveal the successes and failures of the plan. Fewer Oregonians are uninsured than ten years ago, but universal coverage has not been achieved. Medical care costs are rising slower but continue to cause havoc in the state's budget. In September 2000, the Governor convened a Summit of health leaders to regenerate bi-partisan support and commitment. The promise of solidarity, community involvement, and bi-partisan collaboration will determine future success of the drive toward universal coverage in a cost-controlling system.

1. INTRODUCTION

In June 1989 the President of the Oregon Senate, John Kitzhaber, M.D., gave the commencement address at the Oregon Health Sciences University graduation. He lamented the fact that the dynamics of health care financing

Changing Health Care Systems from Ethical, Economic, and Cross Cultural Perspectives,
edited by Loewy and Loewy. Kluwer Academic/Plenum Publishers, New York, 2001.

in the previous decade had generated a spiral in which the costs of health care and the number of persons without health insurance were both increasing. As he addressed these newly minted dentists, nurses, and physicians, the bills making up the Kitzhaber-designed Oregon Health Plan (OHP) were about to pass into law. He warned, "This society will not long tolerate a health care delivery system that excludes 40 million Americans...."[1] Ten years later, in January 2000, Governor John Kitzhaber, in his state of the state address, called for renewed commitment to the social goal of the Oregon Health Plan. He took pride in the fact that the percentage of Oregonians without health insurance had fallen from 17% to 11%. Again, he lamented, "Yet, in spite of that progress, one out of ten Oregonians, more than 300,000 people, are still without health insurance coverage—more than 66,000 of them are children. That is simply indefensible."[2]

In this paper I offer an interpretation from the point of view of social ethics of the process that occurred during those ten years. I identify the core ethical questions imbedded in the Oregon Health Plan's history. I explore key the political solution on which the OHP rests: a blending of public and private systems. I trace the ethical ground for the key feature of the OHP, its list of health service priorities. Finally, I offer an assessment of achievements and continuing tasks of the Oregon Health Plan.

2. THE SOCIAL ETHICS FRAME OF REFERENCE

At the heart of the social motivation of the OHP lies the question of the moral purpose of society's Medicaid commitment. I submit this is a response in solidarity to those with health needs whose income level prevents access to health care. This is a form of social solidarity. It is an expression of expectations that membership in a community creates a common good in which all should enjoy a fair share. The first ethical element of the OHP is this grounding in solidarity. It was evident in an expression often repeated in debates about the OHP: it would "ration services, not people." A driving vision for the OHP was and is the idea that universal coverage is a common good that all citizens should enjoy.

In addition to the basic moral commitment of solidarity, the OHP also sought to create a more rational, prudent, and wise pattern of resource allocation. Health care dollars are scarce in public and private budgets. The intense rate of growth of health care spending in the decade leading up the creation of the OHP set the stage for the way solidarity with the poor had to be expressed. It had to be associated with a braking system that could maximize the number of persons with coverage for important health services.

The OHP sought to work out that balancing act through the mechanism of a list of prioritized health services. The most important items on the list would be protected by a prior decision to cut services systematically from the least important items. The priorities were to be an expression of community values combined with expert information about probabilities of success of various treatments. The prioritized list was to function as a statement of the relative value of various health services. Deciding about the relative value of services is the responsibility of the Health Services Commission, a new commission created by the original OHP law.

Thirdly, the OHP sought to produce a fair system of sharing the burden of getting to universal coverage. Initially this took the form of a Medicaid commitment that cut off at the Federal Poverty Line, after which employment-based insurance was to take over. Statistics showed that the vast majority of the uninsured had incomes above the poverty level and were either employees or the dependents of employed persons. The strategy of a mandate on all Oregon employers to provide health insurance for their employees (with employee contribution not to exceed 25% of the premium) was the first idea for creating a fair solution to the distribution of financial burden. The mandate fell victim to existing federal protections of employee benefits and was finally abandoned in 1996. The second idea reached legislative form in 1998 and consisted of subsidizing premiums for individuals with incomes above the poverty level.[3]

Finally, the OHP looked beyond health care itself and acknowledged that health care is but one of the goods government pursues with collective dollars. Responsibility for the distribution of available dollars among socially useful expenditures remained with the legislative assembly. It is the responsibility of elected officials to make decisions about maximizing the quality of life in community by investing in a mix of health care, education, economic development, corrections, transportation, and other services that benefit citizens.[4]

3. STRUCTURE OF THE POLITICAL SOLUTION: PUBLIC PRIVATE PARTNERSHIP

In 1989 as in 2000, the vast majority of persons without health insurance have incomes near the Federal Poverty Level. In 1989, Oregon's Medicaid program was tied to the level at which the state gave financial aid to families with dependent children (AFDC). Eligibility for Oregon Families had fallen to 58% of the poverty level. If this level were still in place in 2000, a person with an annual income of $4,850 would not be poor enough to qualify for Medicaid. Part of the solution was to raise eligibility levels to 100% of the federal poverty level ($8,350 annual income in 2000). The other part of the solution was to stimulate the health insurance market so that uninsured persons above poverty would be more likely to be offered health insurance through their place of employment.

The Medicaid reform component of the Oregon Health Plan has three social ethics elements. First, clarify the moral purpose of Medicaid as solidarity with the poor by extending eligibility all the way to the Federal Poverty Level (rather than stopping at 58% of that income level). Part of the clarification was the elimination of categorical eligibility (such as having dependent children, or being pregnant).

Secondly, the reform called for defining a benefit package based on a prioritized list so that the most important and useful services would be protected from budgetary cutbacks. The list was to be developed with community input about values and expert information about the probability that a given health service (e.g., medicine for high blood pressure) could deliver the valued outcome (e.g., prevention of stroke).

Thirdly, the Oregon Health Plan promoted solidarity by identifying the Medicaid benefit package as the benchmark benefit package which would be used to determine whether or not a private insurance package qualifies as meeting the obligation of mandated insurance for employees. This element sought to tie the interests of the middle class to the interests of the poor. That is, the best way for the middle class to assure that they had adequate health care packages through their places of employment was to be sure that the Medicaid package was fully adequate. The wedding of interests was intended to transform middle class consumerism into an inclusive concern. The quality of Medicaid would be the benchmark for judging the quality of private insurance.

The requirement to manage the Medicaid budget was clear. The legislature should use the list to make budgetary adjustment (funding fewer services if needed, or expanding services in an orderly way if financing permitted) The reform also called for general system efficiencies (along with the line adjustments) through the preferred use of capitated managed care to organize the delivery of services. A third aspect of the budget management strategy was the use of a dedicated cigarette tax to pay part of the costs of expanding eligibility up to the poverty line.

The other half of the Oregon Health Plan is the private insurance reform effort. First, legislation imposed a mandate on all employers to participate in the purchase of insurance for their employees. The mandate was conditional. If employers achieved stated goals voluntarily, the mandate would disappear. If the goals were not met, employers would either participate in the purchase of health insurance or pay a tax equivalent to the costs they would have incurred. The mandate strategy ultimately failed politically and the legislature formally repealed it in 1996.

The Oregon Health Plan insurance component also includes several other reform measures. A state agency supervised a low cost insurance product for small businesses. The state organized a high-risk pool involving major carriers for persons who were unable to get coverage in the usual market. A special task force created an insurance product that was substantially identical to the Medicaid benefit package (1991). Anyone seeking to purchase this plan could not be refused coverage or charged higher premiums because of health status. Further reforms in 1995 created equitable prices between large and small business seeking group coverage and required carriers to provide affordable continuation plans for persons leaving employment through which they had coverage. In 1997, the legislature passed the Family Health Insurance Assistance Program that provides subsidies to households whose incomes fall below 175% of the Federal Poverty Line. Funding decisions by the legislature limited this program to a fraction of potentially eligible households. In addition, the legislature set up a state Children's Health Insurance Program to provide insurance to currently uninsured children up to age 18.

In Oregon the Medicaid program serves an average of 84,000 persons each month who would not be eligible under the traditional Medicaid program (which serves 268,000) persons. An estimated 27,000 persons potentially eligible for Medicaid remain outside the system. The private insurance components of the Oregon Health Plan directly serve fewer persons. The Children's Health Insurance Program reaches 16,000 children, and the Family health Insurance Assistance Program reaches 6,000 persons. The high-risk pool serves 5,700 persons. The Family Health Insurance Assistance Program serves 4,000 persons with a waiting list of 26,000 potential applicants.[5]

4. A SOCIAL ETHICS ASSESSMENT

The original vision for the Oregon Health Plan was universal coverage. Still Oregon has "more than 300,000 people" without coverage as Governor Kitzhaber reminded the citizenry in his 2000 State of the State speech.[2] Only 10% of this group would be eligible for Medicaid. The most profound challenge for the Oregon Health Plan lies in shortfalls of the private insurance components of the plan. How has the Oregon Health Plan performed from the point of view of social ethics concerns for the common good, fairness, prudence, and wisdom?

4.1 The Common Good

A social ethic concerns itself with the quality of life of the community. By *common good*, I mean the benefits that come from life together in community. I mean the benefits that support human flourishing of all individuals in the community. True common goods flow into the wellbeing of all members of the community. The common good is an ideal toward which historical societies strive. The common good defines a community's quality of life. It is more spiritual than material.[6] In the context of health care, the common good is not the services provided, but the security of access to aid in a time of need. The actual services provided are private, individual goods that only a few persons need. The social solidarity in compassion that binds all members of the community to each other is the common good.[7]

There are three positive and two negative evaluations of the Oregon Health Plan from the point of view of the common good. From its origins, the plan featured a commitment to civic involvement in its workings. The law required the Health Services Commission[8] to convene community meetings to ascertain the community values that should guide the priority setting process for health services. A civic group, Oregon Health Decisions convened community meetings throughout the state in 1990. The Commission combined the identified community values with technical data from health care experts to construct the list. This structure and practice of community involvement has continued throughout the ten years of the OHP and nourishes the sense of community responsibility for health care.[9]

The second positive feature seen from the common good perspective is the fact that the state has made a formal commitment of access for all the health care. The commitment itself becomes part of the Oregon common good, even while the actual achievement of the social mechanism falls short of the goal. The commitment and momentum enhance the common good.

A third element of common good associated with the Oregon Health Plan comes from the fact that the prioritized list integrates all health services into a unified, integrated frame of reference. The plan sees dental health, mental health, and health of the rest of the body as the object of health care. This integration provides an enriched conceptual scheme for planning and organizing the services that correspond to the common good of health-related security and welfare.

On the negative side of the common good account, the Oregon Health Plan has fostered dangerous illusions about the financing of health care.[10] Confusing beliefs about the role of employers in the distribution of health insurance continue to cloud the discourse about solutions. These confusions are not unique to Oregon. Their continuation in Oregon, however, hinders progress toward achieving health care access as a common good of Oregonians.

The mandate strategy sought to impose a legal obligation on all employers. Many advocates assume that employers have a moral obligation to provide health insurance. The problem is that language about employer contribution hides the fact that benefits are part of total compensation determined by the market for labor. Households pay insurance premiums. Households have the responsibility of contributing to the pooled resources that provide coverage. The illusion that employers pay for insurance and give it to employees stands in the way of community awareness of social responsibility.[10] The use of a portion of a tobacco tax dedicated to the health plan further hides the general societal obligation. It shifts attention away from the whole community to those "sinners" whose smoking makes them convenient targets for imposing an increased burden for financing health care. This strategy fails to include everyone in the duty to share the burden of the common good.

A second negative aspect from the point of view of the common good is the breakdown of bipartisan support for the Oregon Health Plan. The emergence of the plan's initial legislation rested on remarkable bipartisan support. This support has slowly degenerated over the decade. As shared vision and commitment to the common good goals of the plan diminished, the struggle to secure sufficient funding both for Medicaid and the Family Health Insurance Assistance Program has come to be seen in terms of partisan wins and losses. The partisan frame hides the question that a common good loss to the whole Oregon community may be the result of certain "victories." When Governor Kitzhaber convened a broad bi-partisan group for the Summit on the Oregon Health Plan, a central purpose was "to rebuild trust" and to reinsert a bipartisan spirit into the work ahead.[11]

4.2 Fairness

A significant achievement from the perspective of fairness lies in using the idea of poverty as the primary criterion for Medicaid eligibility. The plan invoked the insight that the primary reason for reaching out to offer health coverage to persons with low incomes ought simply to be their poverty, not marital or family status. This insight drives the innovative aspects of the Oregon Medicaid program, although categorical programs continue to exist side-by-side with the new approach. This duality continues because the Oregon Health Plan Medicaid component is a demonstration program operating under the conditions of a waiver of the usual federal Medicaid rules.

The focus on poverty as a reason for assistance also informs the Family Health Insurance Assistance Program. The central feature of this program is income-based subsidy of health insurance premiums. The program focuses on those just above the federal poverty level (up to 175% of FPL). Lack of insurance as a social phenomenon concentrates in these lower income levels. Sixty-two per cent of Oregon's uninsured persons have incomes below 200% of the Federal poverty level.[12] The major reason for going without insurance at these income levels is affordability. The idea of fairness fully supports this policy of seeking equity in the degree of burden faced by persons with low income as they participate in the common good of access to health care. The subsidy comes from state general funds collected mainly by income taxes. Thus, this approach spreads the burden of health insurance for the near poor through the entire range of income levels.

Negatively, from the point of view of fairness, the Family Health Insurance Assistance Program has incorporated an unfair feature. Focusing on the goal of getting families without insurance into coverage, policy makers made being uninsured for six months a condition for eligibility. This means that families who had sacrificed to purchase health insurance would not be eligible for the subsidy (unless they went without health insurance for a period of six months). However, the reason for the subsidy is income level. The criterion violates the idea of fairness.

A second negative mark from the point of view of fairness lies in the complex relationships among health care providers. The Oregon Health Plan seeks to purchase Medicaid services through contracts with managed care organizations on a capitation basis. One of the enduring problems with market competition in the health care arena is the issue of adjusting for adverse selection. One plan might accumulate an unusually high proportion of costly members yet the capitation rate remains identical to other plans with lower concentrations of high cost patients. There will be an incentive to seek ways to reduce the number of high cost members. Unless the effect of adverse selection is smoothed out through cooperative adjustments (risk adjustments), the general effect is for all plans to seek to minimize the number of high cost members.

Participating plans have engaged in discussions about developing a risk adjustment system, but have not taken effective action to overcome the problem. In Oregon, one managed care plan (CareOregon) was created among safety net providers specifically to permit them to participate in the Oregon Health Plan. This plan by definition attracts a high concentration of high cost members and has no offsetting private insurance business to act as a financial buffer. The consequence is a continuing complicated form of unfairness rooted in the competitive relationships among health plans. At the Summit on the OHP, Governor Kitzhaber put this problem at the head of his list of causes of instability that need to be resolved.[11]

4.3 Prudence

As a social ethics concept, prudence looks at the immediate future and asks what needs to be done to have adequate resources to move forward toward chosen goals. Prudence is concerned with strategies for forward movement, supply lines to sustain efforts, staging and timing of activities. The measure of prudence is always the social goal at which the community aims.

In a positive vein, the Oregon Health Plan deserves good marks for its conscious effort to construct a clear plan for cost control of the Medicaid budget. The use of the prioritized list to protect the most important services from budgetary cuts is a prudent plan. It recognizes that resources are always limited and that trade-offs are part of the life of public service programs. The list is an instrument of prudent public policy.

Negatively, prudence has not been well served in the working out of government budget decisions. In the area of Medicaid, legislators have regularly budgeted insufficient funds to cover all of the persons potentially eligible for coverage. This means that the program administrators have been motivated to tolerate less than vigorous outreach to the population of potential clients. Similarly, legislators provided funding for only a fraction of households potentially eligible for the Family Health Insurance Assistance Program. With only enough funding for 4,000 persons, a waiting list of more than 20,000 persons, and a potentially eligible population more than ten times larger, the program is vastly under financed. This leads to the conclusion that the question of prudent assembly of resources for this program has not been sufficiently addressed by legislative leaders. Since this program is the principal vehicle for reaching the near poor households that now lack insurance coverage, the low level of financial commitment to the program is a glaring deficit in social prudence at work in the implementation of the Oregon Health Plan.

4.4 Wisdom

Social wisdom focuses on the question of what the community values most deeply and seeks to clarify social goals in terms of those values. The central goals of the Oregon Health Plan are universal coverage, affordable health care, a rationally designed set of service benefits, and a fair distribution of the financial burden.

Positively, the prioritized list of health services is an important instrument of wisdom. It offers a method for defining a health services benefit package that builds rationally on both community values and expert information. The requirement to be explicit about community values and the use of public participation activities give the OHP a structure and a history that support the expectation of continuing work on clarifying social goals. In addition to the original community meetings conducted to lay the groundwork for the prioritized list, OHP leaders have commissioned Oregon Health Decisions to conduct five additional public participation projects to generate citizen input to the ongoing guidance of the plan.[9]

A second element of the plan that fits the pattern of social wisdom is the development of a benchmark benefit package to determine which packages merit support in state financed programs. Because policy makers ultimately had to abandon the employer mandate strategy, this work on a standard benefit plan has not become normative. Still, it sets a pattern and establishes a method for equity that leaders can use for future development.

The commitment to building the program using both public and private sector elements is a third positive element of social wisdom. Many persons believe that universal coverage requires government to take over a significant portion of the health care system. Throughout the period of development and implementation of the Oregon Health Plan, advocates for a single payer solution consistently urged their case in public forums. Many people hold that a single payer solution is more rational and efficient. They have, nonetheless, supported the plan because it offered movement toward greater access for the poor. There is pragmatic wisdom in seeking to make a program succeed because it heads in the preferred direction. Crafting solutions that involve both public and private sector elements continues to show the most promise for movement in the direction of universal coverage.

Negatively, it is unlikely that the adoption of managed care as the preferred vehicle for achieving significant cost savings for Medicaid held a great deal of promise. It was socially unwise to expect this strategy to yield relatively short term cost savings. Studies that made it appear that large savings would come from managed care had compared highly managed and mature HMOs against unrestrained fee for service. By the time the Oregon Health Plan was being formed, the health care delivery system was filled with a host of new HMO and other managed care experiments. The Oregon Plan further stimulated the formation of loosely defined managed care entities. This phenomenon radically altered the likelihood of large cost savings from Oregon Medicaid's preference for managed care delivery systems. Some argue the point that costs would have increased even more dramatically without the use of managed care. Others remain unconvinced that the central effect of managed care has been its control of costs.[13]

5. CONCLUSION

This review of the experience of the Oregon Health Plan's first decade leads to five conclusions. These apply primarily to the Oregon context, but are relevant also to other communities that share the problem of uninsured citizens.

First, it is important to keep ethical vision on the table as part of the political discourse about health care reform. From the beginning, the Oregon Health Plan was presented as flowing from an obligation binding the state government to assure access to health care for all its citizens. It is important to keep this perspective in the political discourse and not let it disappear under the pressure of technical concepts needed for economic and organizational problem solving. It is important to keep the ethical rhetoric honest by taking its challenges seriously. Appeals to community, fairness, prudence, and wisdom too easily become hollow rhetoric that feeds cynicism. In his address to participants at the Summit on the Oregon Health Plan that he convened in September 2000, Governor Kitzhaber set the continuing policy problem in terms of the core social values underlying the OHP.[11]

Second, policy leaders should work diligently to maintain, nurture, and renew bipartisan support for completing this unfinished task. Partisan victories with winners and losers that leave the common good injured are losses for both sides. Bipartisan collaboration is the exception rather than the norm in politics as usual. This quest of universal coverage has to break out of that usual pattern, and then break out again when the old habits take over.

Third, maintaining a relatively high level of civic involvement is important for success. This is not because ideas are lacking to the leaders. It is rather that political will and commitment in this matter depend on willingness to share burdens as well as benefits. Public dialogues can engage citizens in questions about their common good. Opportunities can and should be sought by leaders to promote these dialogues.

The latest effort in this vein, *Making Health Policy 2000*,[9] combined a telephone survey, community meetings, and focus groups to explore Oregonians' values about the health care system. This project, conducted by Oregon Health Decisions, created public input for several state agencies and commissions in preparation for the Governor's Summit.

Fourth, as long as the state leaders prefer to build toward universal coverage with private and public sector approaches, efforts to make those elements fit together smoothly are crucial. Medicaid leaders should be wary of the ultimately self-defeating effects of surreptitiously shifting costs from public budgets to private sector budgets. Cost control strategies in particular should be closely examined for subtle cost shifting practices that give only the illusion of cost control without effectively accomplishing the necessary tasks that social prudence and wisdom require. Public leaders should work hard to keep private sector health care organizations willing and collaborative participants in creating and maintaining smooth transition points between Medicaid and private insurance.

The Oregon Health Plan now has ten years of history from which to learn. Its once highly controversial prioritized list of health services is no longer shocking.[14] The outrage about rationing care has quieted with the recognition that rationing goes on in every jurisdiction, just not openly and explicitly. Still, three hundred thousand Oregonians remain without health insurance and medical care inflation continues to wreak budgetary havoc.

In his Summit speech,[10] the Governor set the challenge for the next stage of policy development for the OHP. He identified four crucial commitments required to achieve progress:

1. Maintaining a priority-based approach to health services as benefits
2. Recommitment to social responsibility for caring for the poor
3. Recommitment to universal coverage
4. Achieving an effective control of medical inflation

Achieving the primary goal of the OHP remains in the future. The ethical vision of health care as a common good still requires political commitment to fairness, prudence, and wisdom to carry the state's community to the desired goal. Getting there will not just happen. It is a work of the people and their leaders. It is a clear example of how ethics and politics must come together to express the meaning of membership in a community.

ENDNOTES AND REFERENCES

1. Oregon Health Sciences University, *Views*, Summer, 1989: 22-23.
2. http://www.governor.state.or.us/governor/speeches/s000121.html.
3. See the description of the Family Health Insurance Assistance Program at http://www.ipgb.state.or.us/Docs/fhiaphome.htm
4. See M. Garland, "Oregon's Contribution to Defining Adequate Health Care," *Health Care Reform: A Human Rights Approach,* ed. by A. Chapman, 1994, 211-32. See also M. Garland, "Justice, Politics and Community: Expanding Access and Rationing Health Services in Oregon," *Law, Medicine and Health Care* 20: 1-2 (Spring - Summer 1992): 67-81.
5. Enrollment data provided by Oregon Medical Assistance Programs and Insurance Pool Governing Board.
6. See the classic treatment of this concept by Jaques Maritain, *The Person and the Common Good,* 1947, 52-53.
7. See Michael Walzer's discussion of medical care as a component of security and welfare in *Spheres of Justice*, 1983.

8. H. Klevit, et al., "Prioritization of Health Care Services, A Progress Report by the Oregon Health Services Commission" *Archives of Internal Medicine* 1991;151: 912-916.
9. Reports from these projects are available from Oregon Health Decisions at Box 125, Tualatin, OR 97062: "A Common Purpose in Health Policy," 1994; "A Common Voice for Health Reform," 1994; "Consumers Want Choice and Voice," *Grading Health Care*, ed. by P. Hanes and M.R. Greenlick, 1998; "Searching for Fairness," 1998; "Making Health Policy 2000," 2000 (available also at http://www.ahd.org/states/OR).
10. Uwe Reinhardt, "Reorganising the Financial Flows in U.S. Health Care," *Health Affairs* vol. 12 (1993), Supplement, 172-193.
11. See http://www.governor.state.or.us/governor/speeches/s001013.html.
12. See data on the uninsured in Oregon at http://www.ohppr.state.or.us/faq/povest.htm.
13. See Kip Sullivan, "On the 'Efficiency' of Managed Care Plans," *Health Affairs*, 19,4 (2000): 139.
14. See, for example, Peter A. Ubel, *Pricing Life: Why It's Time for Health Care Rationing*, 2000.

Chapter 5

A Mortgage on the House of God

The Impact of Managed Care on Community Hospital-Based Graduate Medical Educational Programs

Perry A. Pugno
Director, Division of Medical Education
American Academy of Family Physicians
Leawood, Kansas 66211
e-mail: ppugno@aafp.org

Key words: professional graduate medical education, medical training, professional collaboration, patient-physician relationship, continuity of care, managed care, capitation, outcomes monitoring, clinical problem solving, patient advocacy

Abstract: No one can deny that managed care has had a profound impact on the US health care system, especially in areas of high penetration. Patients are now referred to as "members," "covered lives," or even "units of service." Physicians are simply "providers." Few outside the medical education community, however, are even aware of the impact of managed care on physician training programs, especially those based in community hospitals. The impact of managed care on graduate medical education is multidimensional. That is, it affects many very different aspects of a training program. These include the patient base itself, the fiscal status of the training program, the viability of the sponsoring institution and the resources available to support physician clinical education. In some cases, that impact can be positive. In others it is clearly negative. The priorities of contract populations, clinical guidelines and pathways, capitation, outcomes monitoring and information systems often collide with those of continuity of care, clinical problem solving, patient advocacy and professional collaboration. Significantly, the impact of managed care is also temporally unreliable, in that the patient population of hospitals, clinics and other clinical settings are at risk for major change with minimal advance warning. When coupled with the stresses inherent in residency training, for example, this "instability" of the environment can bring with it substantial challenges for both faculty and housestaff.

Changing Health Care Systems from Ethical, Economic, and Cross Cultural Perspectives, edited by Loewy and Loewy. Kluwer Academic/Plenum Publishers, New York, 2001.

This session seeks to explore some of the aspects of graduate medical training programs most vulnerable to changes driven by managed care, and the ethical challenges those changes can create for both the program participants and the patients they serve.

To begin our discussion of the impact of managed care on community hospital-based graduate medical education (GME) programs, let us first take a brief look at the impact of managed care on health care in general. The first perspective is the new language that managed care has given us. What's in a name? Patients are now being referred to as "members," "covered lives," maybe even "the stroke-protocol patient in Bed No. 3," and the most degrading of all "units of service." The physicians who care for them are simply being referred to as "providers." My analysis of this is that managed care has generated a widget factory mentality, dehumanizing the physicians as well as the patients.

A concern you'll hear me repeat as I move along on this subject is that I believe managed care has created a significant interference with the physician-patient dyad. It has become the new physician-patient-payer triad, basically, the "third person in the room." Now the payer is part of the decision-making process. I believe this compromises the unique relationship of physician and patient that is established over a period of time, based on trust and communication.

The impact of managed care on training programs is not unlike that of its impact on the physician-patient relationship. The patient base, training program fiscal stability, the viability of the training institutions themselves, and the availability of educational resources...all of these elements have suffered a negative impact by managed care in recent years. I hope to provide examples shortly that will clarify some of these impacts.

Let me next review the competing priorities between managed health care and so-called traditional health care. The value system for most of us who went into the business of health care includes continuity of care, clinical problem solving, mission values like care for the underserved, patient advocacy, collaboration among professional colleagues, and an atmosphere of inquiry to try and find ways of doing things better. These are in contrast to the priorities of health care financing in a managed care format, where contract populations are critically evaluated and clinical guidelines are promoted ostensibly because they improve outcomes, but pragmatically because they save costs. Capitation is a method of payment designed to control costs. Outcomes monitoring too often is focused on outcomes that are fiscally driven. Information systems are a valuable tool but, unfortunately, we have lots of information systems to track finances and few that track patient clinical parameters and their outcomes. Finally, there is the priority of profit, the driving force for too many managed care payers. If you contrast these two value systems, I think you can see how this challenges many of the reasons that health care providers gravitated to the helping professions in the first place. I also believe that managed care has inserted the dynamic of "greed" to which some of the other values of managed care, the positive ones like improving patient outcomes, for example, have become subservient.

Last year, our conference convener, Dr. Loewy, prepared a publication that included a brief review of the ethical dangers of managed care. He noted that managed care has provided for externally imposed and economically driven time constraints in the patient-physician interaction. It has created a loss of personal responsibility for clinical decisions when those decisions are guideline driven and economically controlled by payers. Resources available to patients are defined by money rather than by patients' needs. Outcome studies, as I mentioned earlier, emphasize cost rather than the clinical benefit to patients. Finally, there is the concern that managed care has impaired the natural curiosity and stifled the imagination of health care professionals in this country. Once again, the priority seems to be profit driven. I think it is fair to say that the saving of resources, which has been one of the central tenants of managed care, is unfortunately not done so as to provide more equitable care to all those in need, but to increase profit.

Specific examples of how these ethical dangers impact GME programs and community hospitals are what I am going to discuss with you next. In the early 1990s, a primary care training program was established in a setting of heavily penetrated managed care. Despite the recruitment of a nationally recognized physician director, in a period of less than seven years, this training program experienced three transitions of leadership. It was ultimately closed by an action *independent* of the teaching faculty. The justification for such precipitous decision-making was simply economic competition. The impact on trainees, teachers, and the populations they served was considered simply "irrelevant."

Annual population shifts occur as payers and employers contract on an annual basis. Once they get a "better deal," employers simply shift all of their employees to a new insurance plan, which may indeed be based on only very modest cost savings. The impact, of course, is a loss of continuity of care for patients with their physicians since they must go out and locate new physicians "recognized" by the new insurance. For a training program, this can compromise its very existence when it suddenly loses perhaps as much as three-quarters of the patient population.

Medicare—in its role of providing financial support to graduate medical education—has also been part of managed care's impact on training programs. The Medicare HMOs have been, in my estimation, "stealing" graduate medical education money for years. Federal support for GME through

Medicare was included in HMOs' annual allocations (AAPCC) from the beginning. This was money clearly earmarked for teaching programs. Unfortunately, the reality is that these funds were actually passed through to the training programs only rarely. Instead, this money has been feeding the bottom line of many managed care organizations for years. When the legislature discovered in 1997 that the funds for GME were not being passed down to the training programs, they decided to take it away. The resultant outcry by the managed care organizations (MCOs) was so severe that the legislators relented somewhat and decided to take it away from them over a five-year period. Unfortunately, training programs were unable to muster a similar outcry when they lost the money in the first place. Whether that money ever finds its way back to the training programs for which it was originally designed has yet to be seen.

Managed care has also stimulated environments with intense competition, decreasing prices without a similar decrease in cost. Consequently, hospitals have seen their margins decline substantially. When those margins go down, so does the fiscal (and sometimes conceptual) support for educational programs. To cope with this economic challenge, teaching faculties are being pressed to be more "productive" (read: generate more income). Faculty are these days getting tired of hearing the word "productivity." A 1998 Harvard Pilgrim study stated that faculty were 16 percent less productive than other clinicians in a community hospital setting, neglecting the fact that the study didn't take into account the amount of time the faculty was spending teaching trainees. This has driven teaching faculty to spend more time in providing direct patient care service...time that had previously been dedicated to education.

The use of clinical guidelines and pathways has become very popular recently. Unfortunately, I believe they adversely contribute to the present problem of dehumanization of our patients. Patients are put into diagnostic category "boxes." The operating framework is that patients need to either fit into a specific category or the physician is forced to justify why not! Physicians who use these pathways (especially trainees) may find that they are not learning why certain things are done, but only making sure that they are done the way the guidelines say they must be. This clearly stifles the atmosphere of inquiry and attention to individual patient needs that characterize the best training programs.

The next issue I would like to raise is that of the pressure to push patients through our clinical facilities faster and faster. The central priority seems to be time and statistics, rather than making patients better. I believe we are shortchanging our patients and not giving them the human interaction time that they need. The logical extension of some of the decision making that managed care has promoted can, I think, have a substantial negative impact.

For example, I am expecting a new guideline to say that, for patients who show up to the emergency room with chest pain (and maybe even with a documented heart attack), if they have had the chest pain for more than three hours, statistically they are unlikely to die and are therefore not candidates for acute therapy like thrombolysis. Consequently, we should just send them home rather than hospitalize them. I don't think that is the way we should be practicing medicine, and it is also poor role modeling for trainees.

Our elderly patients sometimes require hospitalization with a diagnosis that is necessarily vague. Managed care tends to label vague diagnoses as "unjustified" for admission...resulting in patients not getting the focused attention they need to differentiate organic disease from, for example, depression. How many elderly suffer harm by the delay in clarifying those diagnoses?

Prior to managed care, trainees were taught to seek and manage clinical consultations with colleagues who demonstrated skill and professional behavior. Managed care says you can only consult with whoever is part of the same payer plan. Consequently, physicians are being told they can no longer refer to the colleagues with whom they are familiar and whose work they respect, but must use clinicians of questionable skill who are willing to accept a lower level of reimbursement. Is that what our young physicians should be learning?

The hassle factor inherent in a managed care environment is something that has been somewhat hidden. Recent research has shown that in a heavily penetrated managed care market, primary care providers may be spending as much as a full hour each day in just dealing with the increased paperwork and hassles created by managed care. That is time taken away from what could be spent on learning and teaching. For example, not only has the fixed narrow formulary limited our capacity to teach residents how to use an array of drug resources, but it forces us to waste time switching our patients from one drug to another every time the payer gets a contract for a cheaper brand.

Finally, I am concerned about what kind of role models our teaching faculties are providing to the young physicians in training right now. Our best teachers and role models—should they be spending time being "productive" (just cranking out patients in the clinic) or should they be spending their time with our trainees teaching them how to be competent and conscientious physicians? I think managed care supports an adverse selection toward physicians who are more comfortable with superficial care, the price for which will be paid by the patient in the future. I think this is particularly important in the case of primary care residencies, which are commonly based in community hospitals rather than large academic health centers. Because primary care programs don't generate financial support from surgical procedures, for example, there may not be sufficient income to offset the costs of serving their patient populations. Some of these are underserved and otherwise disenfranchised populations for which the training program may be the only access point to the health care system.

Given those kinds of examples, let's now review what are the outcomes of managed care's impact on graduate medical education training programs in community hospitals. New York hospitals were the first to try eliminating housestaff and replacing them with nurse practitioners, physician assistants and hired physicians. What they found is that hiring mid-levels and contract physicians turned out to be much more expensive than having housestaff and teachers. Some New York hospitals are now in the process of trying to rehire their housestaff and rebuild their training programs.

Training programs themselves are seeing some significant adverse outcomes. The Residency Review Committee (RRC) is an accreditation body that evaluates the suitability of post-graduate physician training programs for a single specialty. An adverse action occurs when an RRC identifies sufficient compromise of a training program's functions to call into question its capacity to meet minimum standards. In one primary care specialty, for example, the previous annual rate for adverse actions has averaged approximately eight percent. In the past 18 months, that committee has increased its adverse action level to nearly 35 percent. Why? Because many training programs have lost sufficient educational resources and institutional support to result in a compromise of their educational missions and capacity to meet all curricular requirements.

I think it is not an unrealistic estimate to say that, because of either accreditation or financial pressures, as many as 10 percent of the United States' GME programs—predominantly those in community hospitals and primary care—are at risk of closing in the next several years. The economic challenges of the University of Pennsylvania and its associated hospitals, for example, put more than 1,000 resident positions in jeopardy last year. The pressures of managed care have also contributed adversely to a loss of leadership continuity within graduate medical education. Not only are we losing teachers but also the directors to lead those programs. David Leach, MD, the Executive Director of the Accreditation Council for Graduate Medical Education, reported at the 1999 Council of Medical Specialty Societies that almost 30 percent of the directors of U.S. training programs left their jobs last year.

Faculties are being likewise challenged to manage an appropriate balance of the service/education priorities. Worries about money and personnel are starting to encroach upon the teaching and role model activities by faculty. A recent journal article indicated that many were willing to change diagnoses and exaggerate physical findings to get services approved to meet their patients' needs. The 1999 Kaiser Family Foundation Study showed that 48 percent of physicians surveyed were "lying" to improve their patients' care. Is that what our physicians in training should be learning?

The end result of all of these challenges is that we are losing training programs, trainees and important access points to the health care system. Many patients cared for by training programs are those underserved and disenfranchised populations of the poor, complex and challenging patients. Some estimate that as many as 40% of the uninsured are cared for by these training programs. These patients are the ones who will pay the highest price for managed care's encroachment upon the viability of GME programs in community hospitals.

Chapter 6

Values in Medicine

Faith T. Fitzgerald
Internist and Professor of Medicine and Assistant Dean of Students
University of California, Davis
Sacramento, California 95817
e-mail: acraffetto@ucdavis.edu

Key words: professional medicine, patient-physician relationship, medical education, medical traditions, managed care, socio-economic determinants of medical care

A survey of the class of 2003—the incoming class at the University of California, Davis—showed that about three-quarters of them had been discouraged by physicians from entering medicine. This is usual among residents and students. Why? Are doctors in such despair that they advise their young not to follow in what was once the dream of their own youth? They are, and they do.

Our generation of physicians "did well" by "doing good." We worked hard but we were "in charge." We were held in high regard, trusted and well paid. By following the advice of our teachers in academic medicine and our mentors in practice, we reaped the rewards to which we felt entitled. And it was not just venality, for among the greatest of those rewards were the confidence and affection of our patients and the exhilaration of discovery. We became embittered when, having created the most remarkable period of scientific progress in the history of medicine, and generated an age where the words "medical miracle" became commonplace, we were suddenly stripped of our authority and stewardship, by as it seemed to us, crass mercantilistic profiteers. Other types of practitioners from New Age gurus to well-trained nurses claimed equal, even greater, value to us. And people who had never

Changing Health Care Systems from Ethical, Economic, and Cross Cultural Perspectives,
edited by Loewy and Loewy. Kluwer Academic/Plenum Publishers, New York, 2001.

practiced medicine, who couldn't spell the conditions for which they rejected coverage, were telling us what to do.

Enraging!

Some doctors quit, and others decided that this, like so many other challenges in medical science, could be mastered by study and hard work. We underwent a series of curricular multifocal myoclonic jerks. We tried to regain dominance of our own patch. Some physicians decided to study business but, as with Proteus, the god who changed form every time one wrestled with him, we haven't quite pinned it down. We thought we could master business as we did "medicine."

Arrogant us.

Worse than anything, perhaps, was the mutation of the doctor-patient relationship: "provider-consumer" just wasn't the same. Patients weren't certain whether we worked for them or for the system. In truth, this estrangement from patients had come sometime before the advent of healthcare reform, as diagnostic technology replaced the touch of physical exam, and subspecialty medicine fractured continuity of care. The science no longer complimented, but now often replaced, the art of medicine. We forgot that the doctor him or herself is a therapeutic instrument. We were fearful of the non-numeric, anecdotal fuzziness of influences on patient illness such as religion, culture, hope, dread, and the primal need of the suffering for magical thought. We shunned the shamanistic side of medicine, and alternative practitioners very quickly wiggled into the gap. And over the past three decades, "wellness" replaced sickness in the public mind as the preferred focus of Medicine, and this "wellness" encompassed emotional and social well being as well as perpetual youthful vigor. Now everybody needed a doctor—not just sick people. And we doctors said: "We can do that!"

Arrogant us again.

So—we had advanced science, developed miraculous technologic diagnostic and therapeutic machines and potions, fractionated patient care among multiple experts, neglected (in fact avoided) our priestly tradition in shame of its intellectual weakness, and tried to expand our hegemony to cover social and emotional disquietude.

Costs soared. Those who paid the bills complained: We were in trouble. Since health was cheaper than sickness, health maintenance was emphasized. Since doctors were trained to care for the sick, others—non-physicians—could often "do health" cheaper and better. Fearful of the competition, we quickly altered housestaff and medical school curricula to cover socioeconomic, political, psychological, cultural, ethical aspects of prevention, screening, nutrition, population medicine; many schools even threw in courses on alternative medicine. All of these new curricular efforts were at the expense of core curricular "sickness" subjects, but achieved no increase in patient contact since, in the more crowded curriculum and the growing need to generate clinical income, teachers had even less time available to listen to and be with students and their patients.

The most efficient and more reliable "virtual" patient, on CD ROM or via actor simulation, was used to teach and examine students while real patients in nursing homes and ward beds complained that doctors spent too little time with them. Sometimes patients were forgotten entirely.

About three years ago the then Dean of my medical school issued a directive to all faculty, residents and students. We would all present ourselves to a two day seminar given by a traveling troupe of for-hire managed care experts. Attendance was mandatory.

- *True*, 80% of our insured were in managed care.
- *True*, increased costs and decreased reimbursements were threatening research, teaching and patient care as faculty were driven to a frenetic pace of clinical work.
- *True*, the jumble of acronyms, payment plans, multiple and mutable regulations, documentation requirements ("magic words", we called them), pharmacy restrictions, and care algorithms were stupefying.
- *True*, we were sending graduating students and residents into the world of mercantile managed care; were we not obliged to prepare them for it?

I phoned the Dean:

"Mandatory?" I said.

"Absolutely. Vital to us all!" He said.

"Everybody? Two days? *All* faculty? *All* housestaff? *All* students?" I said.

"No exception. No excuses. Must do!" He said.

"But who will look after the patients?" I said.

So some of us didn't go.

Now what shall we teach medical students and housestaff? Certainly not how to be the best employees of managed care. Undoubtedly, many of the things in the variably proposed new curricula are important, but frankly dietary, smoking, and seat belt counseling, questionnaires on domestic violence, gun control, and advice on sexual continence and most screening and immunizations can all be done without a medical school education. Think: nurses, physician assistants, and technologists can follow protocols; our previously arcane knowledge is now available on the internet to whomever may seek it; many non-physicians can and do support the patient in his or her search for "wellness." What is so special about doctors? It is a question the managers of for profit healthcare plans often ask, and—answering it themselves—believe that we may be more trouble than we're worth. Why should they pay to create more of us by subsidizing medical education? Even our own accountants at UC Davis referred to the Dean of Students Affairs Office as a "deficit department."

So what can we do that non-physicians can't?

- We can take care of really sick people. *Only* we can do that.
- We can translate knowledge from bench to clinic, from general to particular patients and questions from clinic back to bench better than anyone else can.
- We can generate enormous patient trust if it is clearly the case that our physician's oath—that the care of the sick is our principal purpose—is perceived as real and binding.

We have eroded the public confidence in that oath, in my opinion, by concentrating too hard on the evils of managed care as they affect doctors. We speak with deep feeling of the "uninsured," but not the "uncared for," as if insurance were the only obstacle to our care; we'll give it if we are paid. And we complain about the 7-10 minute visit with complex patients, but we "go along" with it because if we don't *we* will be fired or *our* salaries will be cut. The patients, rushed through their short appointments, wonder if we are making a choice between their well-being and ours. And the students are watching, and learning…this? We talk a lot about "surviving" as professionals in these troubled times, but if in order to survive, we have become what we do not want to be, we may be rightly said not to have survived at all.

Managed care curricula and seminars we attend are replete with advice on coding, organizational structures, efficiency measurements, and the like. In fact, doctors are now advised to become MBAs to succeed in managed care—or to organize into unions to function in it. We have accepted the business model, labor and management, when we should have rejected it from the first. We were not good at it.

In my opinion, teaching health care maintenance and managed care is easy medically and nothing new:

- Know the patient population served and what they need in the way of screening and prevention, always asking whether it does any good and if it is worth the cost
- Diagnose and treat the patients using the best means available and pay as little as possible for the best
- Keep the patient's interest superior to your own

All the rest is business.

Young doctors, like the class of 2003, came into medicine when many, many told them not to. They are vocational. They *need* to be doctors in spite of the absence of our "privileges." We can assassinate their inspiration with our cynicism or, worse, our visible compliance with what we concurrently tell them is bad care and bad science. We *do* have power: Only *we* have the skills and knowledge to care for sick people, and to create new knowledge, and that is what we must teach them. Since students learn by imitation, we must not only insist on caring for real patients in time enough to really care for them, but actively refuse to go along with any system that forces us to do otherwise. Our job as teachers is to teach our students to create a better future for our patients, not how to adjust to a bad one.

Managed care for profit as it is now structured must fail. It is not only dehumanizing (people as "work-units," human tragedy as "medical loss ratio"—it's Orwellian!), but a pyramid scheme economically. Preventive medicine succeeds, but does so by increasing the burden later: It should more properly be called postponative medicine.

And when today's pernicious experiment crumbles as it is already beginning to do, it will be the job of the young doctors whom we have taught to create—with their patients, colleagues and society—a new structure of health care, free of profligacy and entitlements and, most importantly, patient-centered and, I hope, concentrating on sick people.

Environments of practice change. Economic "realities" change. Demographics and social expectations change. Even scientific and clinical "truths" change. But the core value of physicians—to serve their individual patients as best they can with the resources available and to seek for better ways—*that* is the enduring and essential lesson of medical education. If we are denied the opportunity to teach *that*, then we are no longer teaching medicine.

Because such students as I would have us teach are taught to care for their patients with all their devotion, they will be unpopular with management. Not only are they time-consuming and not fully productive, their role models—us—are disruptive to efficient systems when we sustain the weak and disabled, support the socially unvalued patient, and advocate for the disenfranchised patient. Moreover, we must engage in curiosity at the bedside and the bench to set them an example as well as advance the work—all of this drains the commonwealth as it is, a no-profit activity.

Why can't we serve both the sick and society? Simply because their interests conflict, in most cases, and by trying to serve both we would be distrusted by each. Without trust, a doctor is impotent.

Business or the state must supplement medical education because businessmen and citizens get sick, and need us to defend them. But they will only believe we are essential if we act as if we are: So please, let us not be rump economists, nor nurse-practitioners, nor allow ourselves to be lumped together with naturopaths or chiropractors—all "providers" together. We are physicians and should *teach* physicians, requiring that necessary privilege of teaching as a non-negotiable condition of employment.

Our young are the best things we do. What is the future of Medicine? *They* are. Let us not let us, today's doctors sacrifice them in trying to save ourselves.*

*We would like to acknowledge the generosity of the New York Academy of Sciences and thank them for their permission to reprint, in slightly modified form, Dr. Fitzgerald's original presentation to (and subsequent electronic publication by) the Academy. The reader is directed to the following citation:

F. Fitzgerald, "Values in Medicine: How will We Keep Hold of our Oath?" *Medical Education Meets the Marketplace: What Mix of Tradition and Innovation Can We Afford?* Ed by H. M. Greenberg & S. U. Raymond, New York Academy of Sciences, 2000, www.nyas.org/medicaleducation

Chapter 7

Generational Conflicts and their Impact on Thinking about the Healthcare System

Observations of a German Sociologist

Reimer Gronemeyer
Professor of Sociology
University of Gießen
D-35394 Gießen, Germany
e-mail: Reimer Gronemeyer@sowi.uni-giessen.de

Key words: cultural anthropology, family relationships, symbolism, technology, ritual, family conflict, social conflict, health care systems, elder care, allocation of resources, euphemisms, humaneness, quality of life

The people of the Tupinambá were cannibals. They lived along the Atlantic-coast of South America in the 16th century and were victims of the white invaders who promptly eradicated them. Though the Tupinambá were constantly warring with others, they stood under an extraordinary amount of pressure to communicate and harmonize within their own extended family. Ethnologists interpret the Tupinambá's cannibalism as a ritualistic mechanism for relieving these family tensions. How did this come about? What exactly was the procedure? How did they go about it?

Well, first they declared a prisoner of war to be a relative. They then married the prisoner to a woman of the Tupinambá tribe. After having lived peacefully among them for a certain time, he was killed and ceremoniously eaten.

Women held an important role in this ritual: they openly ridiculed and humiliated the "prisoner-declared-relative" who, in return, was allowed to throw little pebbles at them in self-defense. The women were allowed to break out of the duty of keeping the family peace and the man was allowed to take his revenge on the women before his execution. The duty to keep

Changing Health Care Systems from Ethical, Economic, and Cross Cultural Perspectives, edited by Loewy and Loewy. Kluwer Academic/Plenum Publishers, New York, 2001.

peace within the family is stabilized only because the cannibalistic ritual allows them in the end to execute and eat a "relative." Of course, conflicts today between the sexes, groups or generations are no longer solved by means of cannibalism.

Change of setting: I was recently invited to an Austrian television "talkround"—a round-table discussion—on generational conflicts. Around the table sat representatives of unions, employers, state employers, women's rights, scientists and the elderly. The table we were sitting around was floating in a studio high above the city. Glowing in the background, one could see the illuminated Stephansdom, and below it, the glittering lights of the inner-city area. This round table, floating over the town, almost seemed to symbolize the "good old" family table—but perhaps the family table of Europe or Germany differs from the family table of America.

Around such a "family table," most conflicts between generations had been carried out throughout the past centuries. But that sort of table was not a part of our reality. The now present part of the "family" gathered round our table in the television studio was not an association with an underlying feeling of idyllic harmony. The interests here clashed hard and notably. On one side sat the elderly who were expected, finally, to let go of the "steering wheel," which the younger were anxious to grab hold of—not very idyllic, indeed. In 19th century Pommerania (situated in the Eastern Part of the old German Reich), one used to stir for the elderly a so-called "sitting-" or rather "lying down-" powder, arsenic. It was put into the teacups of those elderly thought to be taking too long to "resign." Conflict resolution within families and close knit groups is inevitable—peaceful or not.

Today, (and again, European and American families may differ somewhat) the generational conflict has been transferred from the familiar family circle to the public arena. This changes the "hot" family conflict into a rather cool discussion between lobbyists, mainly on TV. The talkround I mentioned earlier, in typical European television talkshow format, was a setting for individuals who were each presenting their arguments primarily in order to place them in a "better light" politically. The debate between the generations was close to turning into a "cold war," or better, a "pie-raking contest."

A pie-throwing contest is probably well known in the United States, but what do I mean by a "pie-raking contest"? In a "pie-raking contest," everyone, led by their individual interests, tries to clamber the biggest piece of pie onto their plates, hardly a characteristic of the typical "idyllic family table" constellation. The handling of conflicts has been yanked out of its old firmament and moved on to superficial bureaucratic, "higher" levels. Lobbies, representing the elderly as well as the young, take over the regulative functions formerly the responsibility of whole families. In Germany, as well as most of Europe, one can detect a rise of generational conflicts due to the previously mentioned reasons. I will try to elaborate my findings.

As in all industrial nations, the population-structure has changed dramatically. For centuries the populations represented a pyramid-like structure: few elderly, few births and proportionally many children. But the "pyramid" has changed its shape; today, it looks more like an onion. The number of elderly is steadily increasing. The number of 80-year-olds, for example, has increased by 800% within the past 30 years. Italy, an extremely Catholic and, presumably, family-friendly country, has the lowest birth rate in Europe. By the year 2030 the pyramid-now-onion will have turned into a mushroom: The elderly then will make up over 20% of the world population.

At first glance this change of standard-of-living seems rather positive: people who reach the age of 60 today usually seem to be looking forward to a whole new life span. Yet, this is a novel situation. What does it mean for a society to be influenced by an elderly majority? What consequences will we face? How will a "culture of the elderly" change our every-day-lives? Will there be "ghettos" of retirement homes? Will the dominant majority of the elderly over the youth grow and eventually trigger a wave of violence by intimidated youngsters against the "legion" of elderly? What consequences will the inevitable rise in medical costs have on the national budget?

In Germany there are already three inevitable consequences for this kind of demographic development. Economically the younger generation will be bearing the greater part of the resulting "weight." Germany's retirement financing plan requires the working part of the nation to pay into a pension-fund automatically and steadily. This method fulfills its intended purpose so long as the balance between those employed and those retired remains fairly steady. This program fails to function if the proportions of retirees rise while the number of employees drops, as is now already the case.

At the same time, the influence of the elderly in elections is continually growing. Citizens under the age of 18 can not participate in elections. Within a foreseeable time frame, almost every second voter in Germany will be classified as "elderly." So we have a situation where, as the younger generation has to pay ever more into the retirement fund, their political power is waning as that of the older generation is increasing. This phenomenon is not likely to decrease conflicts.

Exacerbating the problem are ecological problems being created by today's generations that will have to be tackled by the following generations. Those living today have merely been shuffling the consequences of their excessive lifestyles onto the next generations; this cannot help but add to the growing conflicts between the generations.

Let me get back to the subject of the erosion of the "old family values." The family had once given a specific role to the elderly: they were respected as grandparents and could also serve as "helping hands" on farms and family businesses. These old roles have disappeared or moved into the background. This raises additional questions of how this last part of a person's life should be filled-out substantively. The financially affluent elderly will spend a good part of their last phase in life as consumers of goods, services and recreational activities. This poorly compensates for the fact that a meaningful and important part of human beings and community has disappeared.

In Germany, the fact that many retirees are financially better off than young families are helps to perpetuate this phenomenon. A possible cultural conflict between the older and the younger generations seems within reach. The elderly in Germany look back on a "classical" family biography as well as a "classical" career. The younger generation now grows up in a society, in which two, once clearly sturdy, structures have been dissolving: They can neither depend on the "traditional" family structures anymore, nor can they depend on the security of the "traditional" professional career. Due this circumstance, any form of communication between the two generations has become increasing difficult. Lifestyles have fundamentally changed.

In Europe (most especially Germany) the elderly have been influenced, predominantly, by a rigid, even authoritative "structuring character," whereas the younger generation has adopted more flexible models of behavior. Making matters worse is a particularly troubled relationship between older and younger generations: the possibility of past Nazi collaboration. Many Germans, when confronted—especially by their children—have continually refused to admit any Nazi involvement. Still, these generations are beginning to die out.

So far, I have been speaking of the difficulties and the conflicts arising out of the obvious over-aging of our societies. I will now try to elaborate some specific consequences for the health care systems.

Germany's health care insurance system has let my generation of Germans grow up in the belief that all people of all ages—i.e., every citizen—deserves and has an unquestionable right to the best possible medical treatment. As I recall, the first heart-transplant challenged that belief: precisely at that moment, for one suddenly realized that medical possibilities were being developed that (a) could not actually be financed for everyone in a population and (b) did not seem applicable to or appropriate for people of all ages.

From then on the question of "rationing medical care," which had not been posed in quite that way before, was on the agenda in Germany. Today, this question is discussed (more) openly. Estimates say that 20% of health care monies is expended in the last months, weeks and days of a person's life. What is also very clear is the fact that every day, the costs for newly developed medical possibilities cannot be financed completely for, and by, everyone. This raises serious questions—moral and otherwise.

At the beginning of the last century, the famous German author Rainer Maria Rilke already vividly described the direction we are moving into now in the collections *of Malte Laurids Brigge*. He writes:

> This exquisite hotel is very old; even in Chlodwig's times one died in a few of its beds. Now one can die in 559 beds. Of course in factory manner [sic]. With such a massive amount of production, the single case (of deaths) isn't undergone very satisfyingly, but what do you expect? The amount is what matters. Who'd care about a well-designed death now-a-days [sic]? No one. Even the wealthy, who could actually afford to die in "a bit more elaborate" style, are beginning to show neglect and impatience; the wish of having ones[sic] own death has become rare. Just a bit longer and it will become as rare as your own life. God! It's all there, isn't it?
>
> You come, find a life and finished: all you need to do is put it on—wear it. You want to leave or have to: well, no big strain: *Voilà! Votre mort, monsieur!* (There you go, Sir! Here's your death!) You die just the way it comes along; you die the death that's part of disease one suffers from (because, since we know all the diseases, we also know of all the different lethal terminations that are a part of theses diseases and not of the people; and the sick actually have no parttake [sic] in it, really...). In those institutions of healing, where one dies with such passion and thankfulness towards doctors and nurses, one dies a death produced by the institution, that's just what we want to see...

You will agree with me, I think, that how a society deals with persons requiring medical treatment is a measure of its humaneness. You also probably remember Aldous Huxley's *Brave New World*, in which people who have gotten "too old" are extinguished systematically in clinics designed specifically as "dying-clinics." How will the medically and technically highest developed societies treat their elderly in the future? How far will one go in order to prolong lives by means of transplantations and implantations of "spare parts"? Do we have to take every possible measure? Should it be left up to the discretion of the individual? The question of guidelines for intervention is tugging at our sleeves more vehemently today than ever before and the intergenerational conflicts that already exist can only exacerbate it.

By now, approximately every second Central-European's life does not end at home. Besides the extremely elderly, the number of those requiring special care is on the rise. Those giving the care are asking for higher wages, because, among other reasons, life is becoming more expensive. In Germany we are predicting a crisis situation due a lack of availability of special care in the not so distant future.

This leads to my assumption that the field of caring for the very old (geriatrics) is heading into a wave of rationalization and automation. Aren't tubes, artificial nourishment and catheters being inserted routinely now because there is such a shortage of specialized caregivers for the elderly? Japanese scientists have recently developed a feeding machine, the thought of which makes me shiver! But: What are the alternatives? What do we prefer: being dependent on a caretaker who hastily tries to feed all of the patients he or she is in charge of or being hooked up to a feeding machine?

Taking care of one's elderly at home, still done in most German families, will be replaced by institutionalized care more and more. And these "institutions" (as we call them in German), in turn, may have to be supported more and more by mechanized "care-robots": VCR-monitoring, computer chips sewn into patients' shirts, etc. Institutionalized care can readily be organized within the private homes. Let me try to outline a scenario for an automated "elderly-suitable" apartment: I'm lying in bed. This bed can move into any wanted position via electronic impulses. It lifts me up and lies me down. My keyboard—within reach—has buttons that initiate my feeding-machine. It makes coffee, bakes bread or softer foods, when my dentures no longer fit. It brings me breakfast in bed and would either serve or feed—depending on my constitution. It can even feed me intravenously—access (for this purpose) having already been implanted into my arm.

It goes without mentioning that my shutters open automatically in the morning and the lights are automatically adjusted according to time of day. A light-switch is a relic of grandfather's times. Of course the entire room-climate (temperature and humidity) is regulated automatically, as well. And after breakfast, the "lavatory-robot" is initiated promptly, after my pressed command: he lifts, cleans and, if necessary, diapers. Before or after the meal, according to choice, he can also wheel me into my bathroom, lift me into the tub, which has automatic water temperature settings as well as outflow mechanisms.

When the feeding-, lavatory-, and washing-robots have done their jobs, it's time for business. I do my finances and shopping online. (The automatic re-stocking of the robots with food and diapers is still being worked on. So far that is still being manually by someone that has to come in the house, but the problem will be solved, soon, I'm sure!). In the mean time the cleaning-robots have sterilized the apartment. The rest of the morning is spent with recreational activities. I read the news in extra-large letters on my monitor—but usually I just watch TV. My CD player is operated via voice-control. And cable connects me with the rest of the world.

Lunch is meticulously put together by the feeding-robot according to best geriatric dietician's recommendations. And in the afternoon: a cozy get-together with old friends and family via conference meeting on the screen.

No problem! Didn't even have to move! In the meantime, an electronic acoustic signal indicates that a bedsore is developing. A turning and pressure-pointless-hanging device ensures that the endangered body-part is rehabilitated.

Before dinner: the body-check. An integrated "medico-robot" checks the bodily functions (blood pressure, temperature, cholesterol level etc.), may prescribe a bit of ultra-violet rays (press 27), a massage or a little gymnastics—each effected by appropriate functions on my bed. It may become necessary to integrate a bit of "some powdery substance" into dinner, in order to lower my blood pressure? The medico-robot will make that decision for me.

A few mind-exercising problems are presented to me on the monitor and, finally, to round things off the "soul-check" or "psyche-check." At the slightest sign of depression the medico-robot involves me into a therapeutic question and answer game. If necessary, a tranquilizer is added to my dinner. This most important of all robots- is connected to the community/ municipal Senior Citizens' Center, which in turn would, if the robot sends the signal that a patient appears to be physically or mentally out of control, immediately send an ambulance. The main ingredient necessary to create such a state of affairs, namely today's senior citizens, are already a part of our reality. How humane is such a vision?

People depending on care will most probably be "better off" as far as the medical and technical possibilities are concerned. At the same time we will become so independent that our relations with others will drastically change and true affection will deteriorate. There's no room for being dependent anymore. The experience of realizing a diminution or cessation of function is papered over by a host of ever increasing artificial possibilities. The conflicts, that once took place in the families, could be avoided and technically by-passed.

A "senior-safe" living environment is constructable, ensuring people their independence and freedom right up to their final heartbeat. At the same time machines, computers and robots could dominate very notably a growing part of the "final phase" of human life…moving into bodies, replacing exhausted hearts, kidneys, hip-bones...

Let's face it: the relics of bodies would be turned into android-like creatures. Merely economic boundaries (seldom moral ones) are distinguishable in this process. Future generation will probably see no sense in having to open the door manually and won't understand that it should be held open for some people who might not have the strength to do it themselves. They will have lost touch with what it feels like to loose strength and become dependent on others. And only fools and the holy will continue to insist that suffering, pain and deterioration are a part of humans and of old age...

CONCLUDING REMARKS

All industrial nations presently converting into post-modern systems are dealing with the same phenomenon: the number of the elderly is on a constant and relative increase. As pleasant as this aspect seems to be to the elderly as obviously are these societies' health systems affected by this to unforeseeable and unpredictable degrees. The "very, very old", tend to be more ill, to need more medical support and treatment and to need more dependent care. Those societies that see themselves as productive societies tend to "marginalize," or even to suppress these elderly citizens and their interests because their needs could be perceived as an unacceptable pressure on the budget.

German health insurance companies are presently liable for the costs, which meanwhile take up over 25% of their budgets. There is growing pressure to enforce rationalizations and cuts in elder care, denying them access based purely on age. They are being asked, more and more bluntly to consider self-financing the cost of most these measures. How closely one might be steering towards an unvoiced desire to euthanize the elderly is not clearly foreseeable at this moment, but nevertheless, the warning sirens should already have begun. The way it responds to the weak, needy and fragile members within its own structures measure the sense of humanity of any society.

BIBLIOGRAPHY

Reimer Gronemeyer, Die Entfernung vom Wolfsrudel: Vom drohenden Krieg der Jungen gegen die Alten. Frankfurt, (Fischer, 1995).

Chapter 8

The Uninsured and the Rationing of Health Care

Diagnosis and Cure in International Perspective

David Chinitz[†] and Avi Israeli[‡]
[†]*Hebrew University-Hadassah School of Public Health*
Jerusalem 91120, Israel
e-mail: chinitz@cc.huji.ac.il
[‡]*Hadassah Medical Organization*
Jerusalem 91120, Israel

Key words: health care systems, rationing of health care, access to health care, government intervention, lack of trust, public discourse, patient-professional relationship

Abstract: This article argues that, when seen in international perspective, the root cause of lack of access to health care in the US appears to be lack of trust in government. Recent developments in the private health insurance sector, however, raise the specter of explicit rationing of health care and, inevitably, summon government intervention in the form of structured public discussion of this difficult issue. Governments in a number of countries have embarked on such national debates, and the experience is reported. Government led public debate of rationing of health care in the private sector may offer a window of opportunity for enacting some form of universal health insurance coverage.

1. INTRODUCTION

Descriptions of the US health system often begin with the statement of a problem—namely, that there is a large and ever swelling segment of the US population lacking health insurance coverage. Taking a medical tack, the United States appears to have a very bad case of "the uninsured, leading to complications of lack of access to health care." However, superior diagnostic

Changing Health Care Systems from Ethical, Economic, and Cross Cultural Perspectives, edited by Loewy and Loewy. Kluwer Academic/Plenum Publishers, New York, 2001.

procedure might begin by asking whether this is the underlying problem or only the most obvious symptom of a more fundamental disorder. In this article, two observers from abroad, but with deep involvement in the US system as well,[1] seek to contribute to this discussion by making a renewed assessment of the problem and suggesting some directions for solutions of the type not usually aired in US health policy debates. Both the problem diagnosis and the potential cures are grounded in the comparative health system approach to which the authors are accustomed. We suggest that the problem of the uninsured is a symptom of a more fundamental pathology: the lack of trust in government. The US is converging with other countries around the need to explicitly ration medical care to the insured population. The emergence of this issue makes experience from other countries, where citizens more readily accept government involvement, more relevant to the US context than is usually assumed to be the case. The evidence from other countries suggests the inevitability of some kind of government-sponsored public discussion of the difficult health care rationing tradeoffs involved. Grappling with this problem has the potential to restore trust in government and in the health system.

2. DIAGNOSING THE PROBLEM: THE HEALTH SYSTEM WARD

Upon entering the health system ward, a visitor would readily note that the United States is not the only sick patient, even among Western countries. The latter also grapple with the difficult dilemma of assuring and maintaining access to health care for their populations while containing costs.[2, 3, 4] The difference is that the United States is clearly the sickest patient in the ward, by comparison with other Western democracies. In fact, those administering care to the other systems often first seek to make sure that none of the egregious pathologies of the US system spread to their own, though they will selectively adopt practices from the US when those seem to work.[5] The US, on the other hand, appears to quarantine itself behind a veil of "exceptionalism,"[6] hardly looking to see if there is anything the other systems might be "doing right," on the presumption that "we are different."

Upon closer inspection, what are the differences and similarities between the US and the other systems?

The first and most obvious difference is that the US has so many uninsured people while most of the other systems have the vast majority, if not all of their citizens covered by some kind of National Health Service or statutory social health insurance program. The second is that the US system, despite leaving so many without coverage, costs the most by almost any measure.[7] Third, the source of health care financing in the United States is mostly private—even though publicly financed programs now account for about half of all health expenditure—while in the countries of Western Europe and Canada the overwhelming majority of finance is public.[6]

Probing deeper, we find that finance of health care in the US has been accompanied by two twin problems not characteristic of other countries: cost shifting and high administrative costs.[7, 8] While various interventions have reduced cost shifting over the last two decades, the result has usually been increases in the ranks of the uninsured, rather than lower costs overall. Moreover, the major cost containment instrument used in recent years, managed care, has created significant public dissatisfaction over increased bureaucratization, lack of choice for patients and physicians, and, at least in highly visible anecdote form, the denial of access to health services. While surveys report increasing dissatisfaction in many OECD health systems, in none does it seem to be accompanied by the ferocious and adversarial tone of the so-called "managed care backlash."[9]

What, on the other hand, are the similarities? Basically, with the exception of the problem of the uninsured, other Western countries share the problems of rising cost, increasing levels of dissatisfaction, and the need to deal with explicit rationing of access to health care services. However, these symptoms are simply not as bad for the other countries. Is it something in the treatments, something about the social "immune" system, or perhaps some kind of behavioral dysfunction that causes large numbers of people to either choose to be without health insurance coverage or be unable to afford it? If all other health systems worked without problems, one might assume that the fundamental problem in the US is the way the system is organized and financed. But it is precisely because other systems are afflicted with similar problems, but seem to manage them better, that the investigator is prompted to look beyond the technicalities of how the system is organized, searching for more underlying explanations about the nature of the underlying pathology.

3. DIAGNOSING THE ROOT CAUSES

Looking deeper, we find what might be closer to the root cause of the problem: American distrust of government and the lack of social solidarity. Findings regarding lack of trust and confidence in government in the US have accumulated impressively over the last twenty years. This was behind the outgoing administration's attempts to restore faith in government.[10]

Other authors take the position, following de Toqueville, that Americans never had faith in government, but they were not unabashed rugged individualists either. What the US used to have but seems to have decayed, is a pre-eminent "civic democracy," activity of voluntary associations which are both based on and engender trust in society which, by spawning creative and synergistic modes of cooperation, leads to economic growth.[11, 12] While a vitally important part of the social fabric, voluntarism and trust based social interaction are fragile and easily corroded by an overwhelming focus on action by the private sector and strong government regulation.[11]

The decline of trust in American society contributes to a heavy emphasis on technical and, perhaps even more so, legal approaches to solving difficult social policy problems. But, according to this literature, such a focus only furthers a vicious cycle in which trust decays. As pointed out by Williamson, in some contexts obtrusive monitoring and sanctioning can be counterproductive to organizational performance,[13] as can adversarial legal proceedings. Many physicians working in managed care environments would probably readily agree.

Those familiar with recent health policy literature and debates in the US would object that the subject of trust is not new, pointing to numerous articles relating to the erosion of trust between physicians and patients, or between patients and health plans.[14] While this is no doubt an important issue and perhaps the crux of much current dissatisfaction in the US health care system, trust at the physician-patient nexus depends on external supports found in the environment. Williamson argues that decisions to engage in transactions without immediate safeguards are due to overarching institutional arrangements that bind parties to an exchange. Arrow argues that the non-profit status of health provider organizations, combined with the certified training and ethical codes, makes it possible for patients to trust that their physicians will provide treatment with an eye to the medical book and not the pocket book.[15]

In the US, cost consciousness has led to corrosion of trust in the US health system despite the type of institutional buffers analyzed by Arrow. The attempt of health plans to "manage care" has often been perceived, rightly or wrongly, as raising the specter of health care rationing even for the relatively well off insured population. The US is moving from implicit rationing through lack of insurance coverage, to explicit rationing for those who are privately insured.[16]

Failure to cope with this problem constitutes a threat to the viability of private employer based insurance, whereas, its successful management presents an opportunity to provide universal access. Other countries have faced the same problem and provided examples of overarching institutional arrangements aimed at rationing health care in a "trustworthy" fashion. Relying on greater trust in their overall political system and systems of civic involvement, other countries are able to introduce cost consciousness, and its logical consequence, rationing, without damaging trust in the health system. They do so through various forms of structured public discussion, inevitably sponsored by national governments, with an eye towards transparency and accountability of the process, ergo, trust.

While the US is unlikely to adopt, wholesale, any of the techniques of public discussion tried in other countries, some form of public discussion will probably be called for to handle the malaise regarding private insurance in the US. If limits to access can be based on accountable, trustworthy mechanisms in the context of private insurance, then the need to limit by excluding large numbers of individuals from health insurance coverage will be reduced, thereby providing better grounds for universal coverage. Thus, the issue of explicit rationing pushes the issue of trust in government, at least as a facilitator of public discussion, into the foreground. Since there does not seem to be any alternative to such government sponsored public discussion, the very process of dealing with this problem will engender greater public trust in government, which could increase the chances for enactment of universal coverage.

4. PRIORITY SETTING AND RATIONING: SOME THEORY AND INTERNATIONAL PERSPECTIVES[17]

Health services have always been rationed. The usual, accepted mechanism for deciding on how services are allocated to patients and populations is reliance on physicians making decisions together with, but more often on behalf of patients.[18] In many systems rationing takes place through creation of waiting lists or the setting of financial barriers. In some countries, implicit, quiet rationing has involved actually denying whole categories of patients access to certain services, even life-saving ones such as dialysis, at the physician-patient level.[19]

Recently, however, the issue of making rationing more transparent and visible has been placed on the policy agenda in a number of countries.[20, 21] The main reason for this is that health systems have become preoccupied with cost-containment, and reducing the scope of services provided is viewed as a high potential method for achieving this goal.

The only problem is that making rationing explicit is very difficult. Attempts to do so strain social consensus to the limit. Indeed, in one sense it can be said at the outset that explicit rationing and priority setting has been a failure, if the criterion for success is a clearly defined and implemented formula for deciding who will get what. Nonetheless, the struggle to find a method of rationing explicitly has produced a number of lessons useful for health systems.

Hall describes well the contributions of various components of the public policy system to the question of rationing health services.[18] Inputs come from physicians and the medical care system itself, bureaucrats such as those responsible for running private or public health insurance programs, expert committees and other sources of technocratic rules, the courts, and democratic political processes. Each sector makes an important contribution, but has significant disadvantages as well. Physicians are closest to the patient and recognize the important unique characteristics of each health situation that should be taken into account in allocating scarce health resources. But physicians usually do not see the larger picture of the needs of the overall population, which may come into conflict with the needs of individual patients. Bureaucrats may operate according to standardized rules that apply uniformly to all cases, but these ignore important aspects of individual cases. Expert committees may suggest rules and guidelines for medical decision making based on epidemiological measures of need and cost effectiveness or cost-benefit analyses. Such rules, however, are based on assumptions about social values that are subject to determination by political processes. Elected political officials are charged in democratic states with articulating the social values that should underlie the allocation of health resources. But politicians are subject to pressures from well-organized interest groups, media reporting of health issues (such as sometimes sensational and melodramatic stories of tragic individual cases). Therefore the political process, on its own, may not reflect the actual preferences and values of the population regarding health services.[18]

Hall and others have concluded, not surprisingly, that there is no magic formula for rationing health services. What seems to be more important is how health systems balance the inputs from the sectors and disciplines mentioned above and come up with a process of rationing rather than a rule for doing it. Unfortunately, this message is overlooked when policy analysts look to other systems for lessons. When it becomes clear that no system has found a formula, it is too readily assumed that developments elsewhere are failures and irrelevant. This is the wrong approach. Perhaps the main lesson to be learned from abroad is not what the specific rules or rationing are, but, rather, how other countries balance statism and pluralism, and technocracy and politics. Review of some international examples will illustrate this point.

4.1 United Kingdom

From 1948 till 1991, the National Health Service in the United Kingdom operated as an integrated command and control system. The main method of allocating resources to health districts, hospitals, and physicians was based on prospective budgeting. In 1991, the British government sought to introduce market mechanisms into the service in order to increase incentives for efficiency. The main element of this reform was known as the "Purchaser/Provider Split." Instead of District Health Authorities directly operating hospitals, they could decide where to purchase services, and hospitals were supposed to compete for contracts with the Districts.

Much has been written about this reform and its outcomes and this subject is outside the realm of this paper. Here we focus on how the reform crystallized the issue of priority setting and rationing. The National Health Service was, and remains, predicated on the notion of guaranteed access to "comprehensive" health services. No specific list of services was ever defined. Prior to the reform, rationing was considered to be a feature of the National Health Service, but it was said to be implicit in the sense that it was based on decisions at the physician level to deny services, often, as mentioned above, on the basis of age, especially in the case of dialysis.

The purchaser-provider split and the emphasis on contracting inevitably raised the issue of defining what is included under the rubric of "comprehensive health services." The National Health Service Executive told districts that they had responsibility for the overall health of their populations and that they should be purchasing for "health gain."[22, 23] As a result, District Health Authorities began to conduct exercises in determining priorities. These ranged from engaging in very technical exercises in cost-benefit analysis to various forms of consultation with the public.[23] In some Districts cases of denial of services—such as the famous case of Child B who was denied a bone marrow transplant on the grounds that the District determined that the chances of success that the procedure would cure her leukemia were too low—received a great deal of public attention and aroused controversy. In this case, and others, court cases ensued and in several instances the authority of the District to deny care based on rational considerations of costs and benefits were upheld.[24]

There are two main messages from the UK case. The first is that Districts rely on a mix of methods to determine priorities, including both consultations with key stakeholders such as members of the public and general practitioners, as well as reliance on cost benefit analysis and assessments of need based on epidemiological studies.[25] Health authorities have not sought to develop explicit lists of covered services. Some districts have excluded or restricted access to specific services such as repeat bypass operations for smokers, or established areas of priority, such as out of hours GP services,[26] but this has occurred only at the "margins."[25]

Nonetheless, rationing decisions by health authorities, and variation across health authorities do occasionally become highly publicized and therefore highly charged. If, for example, access to neo-natal intensive care is guaranteed for babies born at twenty weeks in one health authority, but only at twenty five weeks in a neighboring district, when such disparities are reported in the media pressure is brought to bear on district health authorities. District Health Authorities have in some cases merged, or sought direction from the NHS Executive to set a uniform standard, as a way of reducing the pressure.

Thus, one lesson from the UK is that mixed implicit/explicit rationing can be viable in a setting where the rationing applies to the entire population. However, if different geographic sub-populations within one system are subject to different rationing regimens, the situation may not be stable, and the politicians accountable for the overall system may be called upon to approve (or disapprove) various rationing decisions. This may be one reason why the most recently proposed health reform the UK calls for all services to be purchased by groups of general practitioners. This in effect removes the onus of rationing from both districts, the NHS Executive and national politicians. This accords with the observation that the UK government at the national level has avoided creating a priority setting commission or some other form of conduct of a national discussion of priority setting.[25] Whether or not the rationing role will fall to GPs and whether this is a stable situation remains to be seen. If not, the rationing hot potato may be passed back to the national level. If this occurs, the experience with district level efforts at priority setting, combining technical and public consultative approaches may prove beneficial to the political discussion of rationing at the national level.

4.2 The Netherlands and Sweden

This model, one in which the priority setting process is focused at the national level, underpinned with technical and consultative inputs provides a reasonable description of developments in the Netherlands and Sweden. In the Netherlands a Commission on Priorities in Health Care was appointed by the Minister of Health and published its report in 1992. Known as the Dunning Commission, the body recommended a set of criteria for making explicit choices including exclusion of some services from the public basic basket of health services. The Commission consulted over sixty organizations, and it was estimated that the discussions reached about one third of the population, increasing the awareness and willingness of the population to even consider exclusion of services based on cost considerations. The criteria of the Commission were couched in the language of epidemiology and cost effectiveness: is the service *necessary, effective, and efficient?* An additional criterion is whether or not the service in question is linked to personal responsibility. The Commission, based on these criteria, recommended that homeopathic medicine, in vitro fertilization and dental care for adults should be left out of the basic package. The Dutch Government accepted the recommendations and has used them as a basis for rationing additional services, for example limiting access to physical therapy.[25, 27]

In Sweden a Parliamentary Priorities Commission was created in 1992 and reported in 1995. It also conducted public hearings and came up with a general set of criteria for prioritizing health services. Priority was assigned to treatment of life threatening illnesses, palliative care, and preventive services, in that order.[28]

4.3 New Zealand

New Zealand, like the UK, has built its health reform on a purchaser-provider split. Regional health authorities have responsibility for purchasing care on behalf of their populations. As in the UK, this setup has focused interest on determining priorities that can guide the hand of the purchasers. A National Health Committee was appointed in 1992 to advise the government on health priorities. The Committee decided that it would not adopt an exclusionary list, but, rather, focus on the conditions under which services should be provided. Services were evaluated according to criteria such as benefit, value for money, fairness and values of communities. The Committee worked with expert clinicians, service users and patient groups, researchers, purchasers, and providers. While no decisions to deny service can be directly traced to the Committee's recommendations (and, indeed, at least one highly publicized failed effort to deny dialysis to an individual who did not meet the guidelines suggested by the Committee), its work appears to have had a more indirect effect. Prescriptions of expensive treatments for high blood pressure were reduced in the wake of the Committee's assessment that they were of low cost effectiveness. General public awareness of the need to set limits and determine priorities appears to have increased due to the very functioning of the Committee.[29, 30]

4.4 Israel

Israel may provide, somewhat unexpectedly, the example with most relevance to the US, since its health system is based on a form of regulated competition among HMO type organizations. Since 1995 all citizens are covered by National Health Insurance (NHI) provided by four competing sick funds, operating much like Health Maintenance Organizations. NHI mandates freedom of citizens to enroll in the sick fund of their choice, sets a global limit to sick fund budgets, and mandates a standard benefits package which must be provided by all sick funds. The updating of this standard basket has become a focal point of the Israeli health policy debate. In 1997, after a tumultuous debate over the national health budget, it was decided to allocate about a one-percent increase each year for integration of new medical technologies, especially new pharmaceuticals, into the standard basket. A public committee, including representatives of government, the sick funds, the Israel Medical Association and public representatives, was appointed to allocate this budgetary increment. The process involves a combination of technology assessment and interest group pressure, as representatives of various classes of patients seek NHI coverage for new treatments relevant to their ailments. Not all candidates for adoption are included, as the budget is limited. The committee's activities have received increasing media coverage in each of the three years it has met. While some controversy will always attach to a process like this it seems to have gained enough public legitimacy and accountability to enable it to continue to explicitly ration scarce medical resources.

4.5 Summary

A number of countries have embarked on publicly visible efforts to set priorities and ration health services explicitly. The efforts typically involve a combination of technical inputs, public consultations and political backing. In most cases the specific effects on rationing have been modest. It appears that the main impact has been to get the issue of priority setting onto the public agenda and to raise public consciousness and perhaps to engender a degree of public consensus that limits to accessibility of health care services are not a taboo subject.

5. DISCUSSION: PRIVATE CRISES, PUBLIC SOLUTIONS, TRUST AND THE ROAD TO UNIVERSAL COVERAGE

What the international experience suggests is that some form of public discussion of rationing of health care is inevitable. The various sectoral stakeholders in every health system have gone through several rounds of trying to pass the hot potato of rationing, only to realize that such blame shifting must give way to a more explicit public debate. Even Britain, despite Prime Minister Blair's penchant for obfuscating difficult dilemmas in the language of "modernization" and the "third way" (what might be termed the "Which Blair Project?") appears unable to avoid open discussion of the subject.

Despite the perceived desire, said to characterize American public opinion, to keep government at arm's length from the health system, a similar dialectic is seen and, arguably, may lead to a similar result. Health plans, followed by individual physicians and physician groups, have taken their turn at being blamed, especially in the media, for rationing health services. Government regulators and legislators have responded by mandating various services and raising the specter of patients' rights bills emphasizing the right of patients to sue health plans. To the extent that such interventions raise costs, however, accountability for the latter is unclear. Legislative or regulatory mandates do not, in and of themselves, ensure financing for any increased costs of compliance. Employers seek to avoid increased costs, but are not interested in making the rationing decisions. Attempts to self extricate from this problem may take the form of so-called "defined contributions," in which employees are given a voucher to purchase health insurance policies of their own choosing. This, however, begs the question of whether variability in the richness of policies based on variability in the distribution of income will be accepted broadly by society or become the focus of the next health finance and delivery system backlash.

This dynamic leads to explicit public discussion of rationing. Already, the subject has been broached by discussions of "medical necessity" and how it should be defined in legislation.[31, 32] Independent review panels are mandated in some states to hear appeals of patients denied access to sought after health services.[33] Medical necessity is a thorny issue regarding existing medical services, *a fortiori* regarding the stream of new medical technologies becoming available at a rapid pace. And it is unclear if definitions of medical necessity and independent review panels are able to pay adequate attention to the cost of health services. Technical, legal, definitions of medical necessity, and mechanisms for appeal and independent review, are a useful input for health policy and regulation, but they need to be supplemented by open discussion of the need for rationing health care. As government has already become implicated in regulating managed care,[34] it would seem natural for it to be the main forum for such discussions and decision-making processes. When accountability for making difficult decisions appears to be falling through the cracks between sectoral stakeholders, the problem eventually bubbles up to government. Morone has presented a form of this argument in documenting how attempts to avoid government intrusion in the health system end up inviting just such intrusion.[35]

In the countries described above governments more readily take accountability for setting limits in health systems. Analysts disagree over how explicit or implicit health care rationing should be, but they converge around the need for explicitly defined accountability mechanisms. For example, processes for making decisions regarding the coverage of specific health services should be transparent and allow key stakeholders adequate opportunity to state their claims. Those denied access to medical benefits should be provided with an explanation of the process and the rationale for decisions, and should have the opportunity to appeal.[36, 37] Services probably should be considered in groups in order to permit tradeoffs to be made. Attention should be paid to the mechanisms by which various services come onto the agenda.[38] And, most important, the public discussion should seek to raise public awareness of the need to make difficult tradeoffs in allocating health resources.

The literature on distributive justice suggests that the greater the degree of accountability characterizing decision-making mechanisms, the more trust they merit in the eyes of the public.[39] Developing accountable mechanisms for rationing employer based privately insured medical care in the US, would, then, increase trust in the health care system. Could this be a step in the direction of universal health insurance? There are a number of reasons to think this might be the case. First of all, the precedent of strong government involvement in the health system, even for privately insured individuals would be set. Second, if employees are willing to trust the government to define the terms of access to medical services, or at least structure the process that would do so, then universal health insurance does not have to be perceived as an open-ended entitlement any more than would private health insurance. Fear of an open-ended entitlement has been one of the barriers to enactment of national health insurance. Finally, the very process of engaging in such a public discussion, provided it is accountable and raises overall levels of trust in the health system might increase social solidarity in general which, in turn, might make the enactment of universal coverage more likely.

6. CONCLUSION

It would be naive to claim that the above scenario, involving explicit rationing leading to increased trust in government and social solidarity which, in turn, would lead to universal coverage, is the most likely or even a possible one in the US context. Suspicion of government, lack of an emphasis on equity and a tradition of failed attempts at enacting national health insurance litter this analysis with obstacles. However, it should be acknowledged that the litany of technical policy responses to repeated crises in the health system is not only wearing thin, but also bringing the question of health care rationing ever more into the open. It is notable that limitations to access to care, and how such limitations are determined has become a touchstone issue not so much for public programs, but precisely in the privately insured sector. It is also becoming more difficult to think of how to solve this problem without some substantial government involvement. Whether this dynamic is linked to the issue of the uninsured will depend, as usual in the pluralistic democracy of the US, on the proclivities of political leaders and the resultants of interest group politics, and on the level of trust and social solidarity in society. These, we have argued, are at the core of the diagnosis of the US health problem, and not the technicalities of finance and delivery that have dominated debate for more than a decade. Perhaps the current crisis in private health insurance, then, opens a window of opportunity for a diagnosis of the underlying causes of lack of coverage in the US and, in turn, a better chance at cure. Desperate times call for desperate measures. On the other hand, a new round of technical, legalistic changes may ensue. In connection with health policy, one is frequently reminded of Churchill's comment that "Americans will always do the right thing, after they have exhausted all the other alternatives."[40]

ENDNOTES

1. Avi Israeli, as Director General of the Hadassah Medical Organization in Israel, is intimately familiar with the US scene through his contact with the US based Hadassah Organization. David Chinitz is spending the second of two year long stints in the US studying the health system, during 1992-93 at Columbia and New York Universities, and currently at the University of California, Berkeley. In addition, both are frequent participants in health policy discussions in Western Europe.

2. RB Saltman, J. Figueras, and C. Sakellarides, eds., *Critical Challenges for Health Care Reform in Europe* (Buckingham, UK), 1998.
3. K. Donelan, RJ Blendon, C. Schoen, *et al*, "The Cost of Health System Change: Public Discontent in Five Nations," *Health Affairs* 1999; 18 (3): 206-216.
4. J. White, *Competing Solutions: American Health Care Proposals and International Experience* (Washington, DC: Brookings), 1995.
5. RG Evans, "Life, Death, Money and Power: The Politics of Health Care Finance," *Health Politics and Policy*, ed. by TJ Litman and LS Robins (Albany, NY: Delmar), 1991: 287-301.
6. V. Rodwin, "American Exceptionalism in the Health Sector: The Advantages of 'Backwardness' in Learning from Abroad," *Medical Care Review*, 1987; 44: 119-54.
7. Organization for Cooperation and Development (OECD), *Comparative Data on Twenty Nine Health Sytems* (Paris: OECD), 1999.
8. S. Woolhandler and DU Himmelstein, "The Deteriorating Administrative Efficiency of the US Health Care System," *NEJM.* 1991; 324: 1253-1258.
9. MA Peterson, "Politics, Misperception or Apropos?" *Journal of Health Politics, Policy and Law.* 1999; 24:873-886.
10. CW Thomas, "Maintaining and Restoring Public Trust in Government Agencies and their Employees," *Administration and Society.* 1998; 30:166-194.
11. F. Fukuyama, *Trust: the Social Virtues and the Creation of Prosperity* (NY: Free Press), 1996.

12. J. Purdy, *For Common Things: Irony, Trust and Commitment in American Today* (NY: 1999.
13. OE Williamson, "Calculativeness, Trust, and Economic Organization, *Journal of Law and Economics*, 1993; 34:453-502.
14. D. Mechanic, "Responses of HMO Medical Directors to Trust Building in Managed Care," *Milbank Quarterly*, 1999; 77: 283.
15. KJ Arrow, "Uncertainty and the Welfare Economics of Medical Care," *American Economic Review*, 1963; 53: 942-971.
16. The most commonly cited case of explicit rationing in the US, the Oregon Health Plan, began exactly the other way around, rationing care for those on public programs. While perhaps containing some useful clues as to process, it cannot be considered an accountable or trustworthy method of rationing for the entire population until applied to those with private health insurance
17. This section is based on previous work of the authors. See D. Chinitz, (1999) "The Basic Basket of Health Services under National Health Insurance, Social Security 54" (Hebrew) and D. Chinitz and A. Israeli, "Rationing: Who and How?" *Journal of the Israel Medical Association*, November, (1999).
18. M. Hall, *Making Medical Decisions* (Oxford: Oxford University Press), 1997.
19. H. Aaron and W. Schwartz, *The Painful Prescription: Rationing Hospital Care* (Washington: Brookings Institution), 1984.
20. C. Ham, "Priority Setting in the National Health Service: Reports from Six Districts," *British Medical Journal* 305 (1993): 435-438.
21. A. Coulter and C. Ham, eds., *The Global Challenge of Health Care Rationing* (Buckingham: Open University Press), 2000.
22. B. Salter, "The Politics of Purchasing in the National Health Service," *Politics and Policy,* 1991 21: 171-184.
23. A. Morga, "Rationing in District Health Authorities," Paper presented at the Third Conference on Strategic Issues in Health Management, St. Andrews, 1998.
24. C. Ham, "Tragic Choices in Health Care: Lessons from the Child B Case," *The Global Challenge of Health Care Rationing*, ed. by A. Coulter and C. Ham (Buckingham: Open University Press), 2000: 107-116.
25. World Health Organization, *European Health Care Reform—Analysis of Current Strategies*, Copenhagen, 1997.
26. C. Bowie, A. Richardson and W. Sykes, "Consulting the Public About Health Service Priorities," *British Medical Journal*, 1995; 311: 1155-1158.
27. B. Eiler, Comments at the First International Conference on Priority Setting in Health Care, Stockholm, 1996.
28. Swedish Parliamentary Priorities Commission, *Priorities in Health Care* (Stockholm: Ministry of Health and Social Affairs), 1995.
29. W. Edgar, "Rationing Health Care in New Zealand—How the Public Has a Say," *The Global Challenge of Health Care Rationing*, ed. by A. Coulter and C. Ham (Buckingham: Open University Press), 2000: 175-191.
30. D. Hadorn, "The New Zealand Priority Criteria Project, Part 1: Overview," *British Medical Journal*, 1997; 314 :131-141.
31. California Healthcare Foundation, *Determining Medical Necessity: Balancing Science, Economics, and Policy*, December/January, 1999-2000.
32. Center for Health Policy, Stanford University, *Medical Necessity*, 1999.
33. G. Dallek and K. Pollitz, *External Review of Health Plan Decisions: an Update* (Menlo Park: Kaiser Family Foundation), 2000.

34. S. Altman, U. Reinhardt and D. Schactman, eds., *Regulating Managed Care: Theory, Practice, and Future Options* (San Francisco: Jossey Bass), 1999.
35. J. Morone, *The Democratic Wish* (Basic Books), 1990.
36. D. Hunter, *Desperately Seeking Solutions: Rationing Health Care* (London: Longman), 1997.
37. N. Daniels, (Accountability for Reasonableness in Private and Public Health Insurance," *The Global Challenge of Health Care Rationing*, ed. by A. Coulter and C. Ham (Buckingham: Open University Press), 2000; 89-106.
38. MK Giacomini, "The *Which*-Hunt: Assembling Health Technologies for Assessment and Rationing," *Journal of Health Policy, Politics, and Law*, 1999; 24: 715-758.
39. TR Tyler and P. Degoey, "Trust in Organizational Authorities: The Influence of Motive Attributions on Willingness to Accept Decisions," *Trust in Organizations*, ed. by RM Kramer and TR Tyler (Thousand Oaks: Sage), 1996.
40. Thanks to Dr John Roark, MD, for re-acquaintance with this quote.

Chapter 9

Application and Implications of Deontology, Utilitarianism, and Pragmatism for Medical Practice

Thor Cornelius
MS 4
University of California, Davis
Sacramento, California 95817
e-mail: tjcornelius@ucdavis.edu

Key words: moral theories, ethical practice, deontology, utilitarianism, pragmatism, instrumentalism, conservation of health care resources, patient-physician relationship, goals of medicine, managed care

Abstract: Ethical theories help to shape both the conscious and subconscious background of our ethical considerations. In what follows I shall discuss the classic deontological and utilitarian theories which have been used in dealing with ethical problems and argue that they are problematic because inflexible. I suggest, as an alternative, an instrumentalist or pragmatic ethic similar to that advocated by John Dewey. Such an ethic, by utilizing a variety of antecedent information and theoretical perspectives, can better serve us not only in dealing with problems today but as a springboard for further development, learning and improvement.

1. INTRODUCTION

As medical health care structures have evolved in the United States, practicing physicians have been increasingly placed in a position in which they must fulfill two conflicting roles: they are to maintain specific duties to each individual patient yet simultaneously conserve resources for large

Changing Health Care Systems from Ethical, Economic, and Cross Cultural Perspectives, edited by Loewy and Loewy. Kluwer Academic/Plenum Publishers, New York, 2001.

groups. This kind of conflict, which is now pervasive in every aspect of the average medical practice pits two major moral theories, Deontology and Utilitarianism, against each other. While a deontologic ethic requires the physician to focus on his or her duty to each individual patient, Utilitarianism requires physicians to take a broader view, trying to obtain the greatest good for the greatest number of people.[1] These expectations often create a direct moral conflict in which the physician cannot simultaneously obey both moral theories. As physicians wrestle with the implications of this conflict, an analysis of these moral theories is required to identify their strengths and weaknesses and to possibly find a new moral ethic that may handle both the societal expectation and the realities of managed care. This evaluation should review ideas of social obligation and expectation as well as a comparison of moral theories based on individual and group populations. An analysis of the implications of these moral assignments will also be done, discussing how different moral theories affect the abilities of physicians. An analysis of a third theory, Pragmatism, will also be conducted in an effort to show how a shift to a new ethic can help resolve these conflicts.

2. DEONTOLOGY

Deontology is most closely associated with the philosopher Immanuel Kant (1724-1804). Simply described, it is the study of duties that people have toward one another. It is based on the idea that each individual has the power to determine his or her own moral law. Understanding this, individuals must then also accept that all people are morally free and that each person assigns his or her own duties. Duties are described as being of two basic types, perfect and imperfect. Duties can also be positive, requiring that a person act in a certain way, or negative, requiring that a person not act in a certain way.

Perfect duties are absolute. They should always be followed and never conflict with one another. They are formed using the categorical imperative, an idea that when assigning duties, each person should create them in such a way that they believe that all people in the world can and should abide by them. This is called the universalizability principle.[2] This system, while creating powerful absolute laws, generally creates duties that are negative (i.e. do not harm), but does not allow for the creation of many positive duties. In effect, perfect duties tell us what *not* to do but do not direct us to positively engage in any particular actions.

Imperfect duties, which include ‘positive’ duties such as beneficence (the duty to help others), are argued for on slightly different grounds. Since we acknowledge that all people are capable of creating their own moral law, others' goals deserve equal respect to our own. This respect implies that we should never treat others simply as a means toward one of our ends, but also acknowledge them as an end in themselves.[3] This creates a universal "realm of ends" where we acknowledge the desires and needs of others in addition to our own. Once we realize and respect this "realm of ends" we become partially obligated to help others achieve their goals. This logic gives us the moral ability to accept certain imperfect, positive duties, such as the duty to help others. While these duties are generally required of people, they are not as absolute as perfect duties.[4]

Another important aspect of the deontologic approach is that it is focused on the individual level. When defining duties, individuals consider the effects of the actions on themselves and their consequences to other individuals. Even when looking at large groups, such as the application of the universalizability principle, judgements are made by considering the impacts on the individual level. Thus, the effects on large populations are considered not as a group, but as the effects on each member of that group.

By understanding these ramifications it is easy to see why many physicians have traditionally held a deontologic ethic, and why society generally expects physicians to follow deontologic moral theory. First, the ethical duties of a doctor become based on straightforward rules that, once determined, apply to every patient that the doctor sees. In addition, since these duties are universalizable, they remain constant through time and may be passed down through the generations. Ethical decisions then become relatively predictable, dictated by observing rules that were created sometime in the past. The absolute nature of these rules is reinforced in medicine by the creation of many different ethical codes beginning with the Hippocratic oath, which is a list of duties to be upheld.[5]

The individual focus of deontology also works to the advantage of physicians and their patients. By focusing on individual duties, requirements and expectations, the doctor becomes the patient's advocate, with the patient's health being the physician's primary concern. This belief system creates a reassuring environment that can foster trust and encourage a positive environment for all individuals involved. In many ways this relationship is considered sacred to the medical profession and is itself considered a duty in many ethical codes in medicine.[6] In fact, some physicians go so far as to state that the role of a physician is to be the patient advocate regardless of costs and social obligation.[7,8]

However, the application of deontologic theory also has some fundamental problems. It's detractors first point to the absolute nature of duties. The rigidity and absolute quality of duties creates an inflexibility that does not allow it to deal with real world situations. Quite simply, perfect duties can and do conflict in the real world. Physicians often face situations where perfect duties such as "do no harm" and "never lie" are not absolute when applied to the real world. Should a physician mildly (consciously or unconsciously) exaggerate findings to insure that a suffering patient gets treatment? What if a patient needs a life-threatening level of morphine for pain relief? These situations occur and rigid absolute duties do not always provide adequate solutions.

The inflexibility of deontology also causes problems in that it does not respond to changes in ethical evaluations through time. This creates friction as the values of society shift. The Hippocratic oath includes the vow never to provide abortions, or deliver any substances that may be considered poisonous.[9] These statements, if not at least somewhat out-of-date, need re-evaluation if they are to keep current with social values. As technology, ideas and values change, ethical considerations should not remain stagnant.

Another criticism of deontologic theory is its focus on the individual. It can be argued that while advocacy is important in medicine, a sense of balance must also be created where equity and fairness are considered in the distribution of limited resources.[10] This effect is significantly magnified when individuals do not directly see the cost of their health care. An example of this is in the often cited Baby K case, where a mother felt it was her physician's duty to stabilize all patients, even if this included resuscitating and stabilizing her anencephalic infant.[11,12] The cost to society, estimated in the hundreds of thousands, was never seen by the mother requesting the services. Yet somewhere, society carries the costs. While society has a duty to shoulder some burdens to care for its members, it certainly does not have the obligation to shoulder any and all burdens that patients want society to bear. A physician needs to have some ability to weigh patient advocacy against some scale of social good.

3. UTILITARIANISM

Utilitarianism is a moral philosophy most closely associated with John Stuart Mill (1806-1873). It is a moral theory that states actions may be measured by the amount of good or utility they do for all the people in a certain group. Actions that provide the most utility are then acted upon. In an ideal world, groups using a utilitarian method would then be able to maximize their utility and happiness, thus benefiting all members. This is also referred to as "the greatest happiness principle."[13]

This theory has several beneficial components. The first is that, depending on what you measure, utility may be quantifiable. In measurable forms utility can be analyzed using a cost/benefit analysis. Costs for an action are added up and subtracted from the benefits. This process is repeated for all the alternatives of a situation and the action with the greatest net benefit is performed. This is very useful when comparing multiple actions and also has the benefit of being reproducible and is easily understood.

Another aspect of this moral theory is that it is only concerned with the actual consequences of actions. Utility is measured by results. This simplifies the mathematics of utility because it is much easier to quantify end results of actions than be concerned with intentions or the substance of the actions themselves.

While this theory is relatively more recent in its application to medicine as a whole it has been used for some time in the arena of public health. In that field, populations are generally treated instead of individuals, and decisions are based on maximizing public good. Vaccinations are examples of utilitarianism being put to practice. The rubella vaccine, for example, is given in childhood to prevent a disease that is relatively benign to the individuals actually receiving the shot. The vaccinations are actually given to hopefully prevent the spread of rubella to pregnant mothers and causing devastating congenital abnormalities to their unborn. Each member of society individually takes a small risk of the side effects of the vaccination to insure that the society as a whole is significantly more healthy.

As might be imagined, Utilitarianism has many proponents, especially in the realms of managed care. In fact, some health care organizations use utility as their primary ethic when constructing the ethical framework for their organizations (although they may not call it by its given name).[14] Groups of insured and limited resources that need to be maximized fit nicely into utilitarian models. This, combined with the claimed quantifiability of Utilitarianism makes the theory attractive to managers of healthcare organizations.

To physicians however, the application of utility is less pleasant. Doctors must now consider not just the consequences of their actions to their patient, but to all patients in the managed care organization. While the new difficulty is an inconvenience, the more disturbing effect of utility is that it requires a break from the traditional patient advocate model of deontology. The patient, in certain circumstances, must simply be told something they hate to hear: the word "no." Worse is the fact that many of these patients will be denied care not because the action is dangerous or impossible, but denied because a cost/benefit analysis did not recognize the advantage of the treatment to the entire group. This dramatically changes the physician-patient relationship. Now, instead of trusting the physician, the patient is put into a position where the physician is considered a barrier to their desired treatment. With the patient's needs superseded by the desires of the group, the physician-patient relationship changes from one of advocacy to one occasionally bordering on conflict.

However, there are several more ominous ramifications of a utilitarian theory. The first problem is that the idea of making individual rights subservient to the needs of the group can be carried to a greater extreme than in the discussion above. Utilitarianism does not prohibit the harming of individuals if it is consistent with maximizing the utility for the majority of the group. If, for example, the group is best served by causing the massive suffering of a few individual members, utilitarian mathematics still generates this action as being a viable alternative. A second problem with the theory is that since it is based on consequences and not intentions of actions, it may encourage the idea that the ends justify the means. These two problems together create problems for distributive justice, since utilitarian methods could be used to justify a range of bad outcomes that allow for significant harm to a few members of a group to marginally improve utility for the larger majority.

Another significant problem with the practical application of utilitarianism is the concept of utility itself. The definition is both obvious and unknown. What is it to maximize utility or happiness? In the field of healthcare utility is not nearly as quantifiable as it proponents would have people believe. While money may be an obvious choice of something to declare as utility, it's applicability is questionable in practice. This is because the purpose of healthcare is to decrease the suffering of its individual members. But the question remains: How much does a unit of suffering cost? The "cost" of suffering varies dramatically, and attempting to place it with a dollar amount is extremely difficult, if not impossible. In addition, with money as the currency of utility, suffering people become nothing more than burdens for the larger group. It seems troubling to have an insurance company, whose sole purpose of existence is to help the sick, consider the caring for the sick strictly as a liability.

Another difficulty with assigning utility is that it may have different values depending on which portion of a large group is being asked. For example, the management team in a for-profit healthcare organization puts a very different value on excess capital (i.e. profit) than the general members of the organization simply because their goals and values are different. Most paying members would probably care very little whether the stockholders made a fortune as long as their healthcare was stable, reliable and within their financial means. However, the management team needs to show that they are not only breaking even, but are making a sizable profit as well. Inability to quantify suffering and disequilibrium in values creates fundamental problems in the realistic application of quantifying utility.

4. DEONTOLOGY OR UTILITARIANISM?

With a limited analysis of both deontology and utilitarianism it becomes clear that while both methods have advantages, neither completely lends itself to application in the real world of medical health care. Deontology has the advantage of focusing on the individual and maintaining an intimate relationship between the physician and patient but does not address the reality of limited resources. Utilitarianism, on the other hand, has a sound theory on dealing with the application of scarce resources and maximizing utility for large groups, but has problematic implications in the real-world identification of utility and the possibility of individual abuse.

When we take a closer look at these problems it is interesting to note that in both theories, the very framework that provides its ability to solve moral problems becomes the foundation of its shortcomings. Deontology depends on an individual focus and the discovery of absolute duties to guide its moral decision making, yet this is the very rigidity that becomes problematic. Utility, in turn, depends on a concept of maximizing group happiness to determine actions, yet the very concept leads to distributive justice. In many respects any new ethic must try to avoid the double edges of these swords.

With neither ethical theory providing a completely satisfactory solution, what are physicians to do? While theoretical musings are adequate for philosophers and scholars, physicians must deal with real ethical problems on a daily basis. There are several options available.

The first solution is to ignore the inherent problems, bury our heads in the sand and use only one of the moral theories described. This method would suffice, and is even advocated by several physicians who are proponents of deontologic theory as stated above. However, with the steady changes in healthcare, it is the rare physician who has the clout to maintain a solely deontologic ethic and simultaneously be able to stay in practice in the realm of managed care. Similarly, physicians who adopt a purely utilitarian ethic, unless they are in the field of public health, will likely find themselves abandoned by their patients as the physician-patient relationship decays and is lost.

Another option is to completely abandon both moral theories. While there are a bounty of other moral theories available, the prevalence of Deontologic and Utilitarian theory used in medicine makes this a very difficult option. Society has expectations of physicians that are not limited to medical knowledge. Society demands that physicians maintain certain aspects of the deontologic ethic, just as the pressure of limited resources forces physicians to acknowledge aspects of a utilitarian ethic. To abandon these theories outright would be a difficult task.

Another option is to haphazardly juggle both theories simultaneously, occasionally applying one theory and then the other. While this is what many physicians likely do in reality, it again has a host of problems. First, the conflicting nature of utilitarianism and deontology makes an unorganized approach inconsistent at best and morally problematic at worst. What decides when a utilitarian ethic is applied or when deontology is followed? Without the consistent application of a moral theory, its application becomes suspect. This is especially true when other factors, such as an economic reward structure exist to consciously or unconsciously sway the feelings of a physician. While some physicians may believe that they are above temptation, economic or otherwise, it is a foolish assumption. With physicians in a world of very real temptations, application of a moral theory needs to be thoughtfully applied in a consistent and organized manner.

5. PRAGMATISM OR INSTRUMENTALISM

Pragmatism is a philosophy that focuses on the relation of theory and practice. That is, it takes into account the growing nature of experience to create a constantly evolving, intelligently derived, moral ethic.[15] It was first discussed by Charles Sanders Peirce (1839-1914) but has had significant contributions made by William James (1842-1910), Oliver Wendell Holmes (1841-1935) and John Dewey (1859-1952). Although "pragmatism" is the name usually given to this school of philosophy, John Dewey for many reasons preferred the term "instrumentalism" as denoting more of what he was really after. For our purposes, however, and because the term "pragmatism" has continued to enjoy wide usage, we shall refer to the entire school as "pragmatism." While pragmatism first appears filled with convoluted rhetoric, it is simply an evolving model of the way people think.[16]

Pragmatists hold that beliefs are nothing more than tools that allow us to effectively interact with the world. They do not believe that there is an "essence" of any belief that allows ideas to be absolutely true or untrue. Ideas and beliefs simply exist to be used to help people interact. With this understanding about beliefs, the questions about moral theories change dramatically. Instead of worrying about whether a certain belief system is abstractly true or false, pragmatic theory evaluates beliefs based on how adequately they allow individuals to interact with the world.[17] In the example of Deontology, a pragmatist would not be concerned with the complicated rationale Kant uses justify his theory, but instead looks at the practical implications of using deontology and ask if it allows people to adequately and fairly solve their problems.

Another advantage of pragmatism is that by using this theory about beliefs, a moral ethic can constantly evolve and improve upon itself to reflect an increase in practical experience. Pragmatic theory is not bound to any rigid framework of beliefs. Any theory can be changed, more fruitfully in a fashion modeling the scientific method. It would also be acknowledged that as our experience grows, changes would continue to occur.

Pragmatism's theory of knowledge and ability to have beliefs evolve over time give it a critical advantage over both Deontology and Utilitarian ethics. It lacks the rigid core framework of the other theories that interferes with their ability to adopt new ways of thinking. It is also applied in a practical way, based on how adequately it allows people to deal with the world.

How would a pragmatic ethic look? A complete analysis is beyond the scope of this paper but it might begin as an evaluation of deontologic and utilitarian belief systems. Once this evaluation was complete the next step would be to acknowledge the beneficial parts of each theory. Pragmatists would then attempt combine the best aspects of both deontology and utilitarian ethics. Once this new belief system is created, it would be put into to practice with the understanding that it would inevitably change and improve over time. Then as experiences grow and new ideas are postulated, the belief system would be further modified. Each change would be made on a conscious level, done in a thoughtful manner with a defined goal. Physicians would have a concrete set of beliefs to follow that could be applied in a consistent and intelligent manner. Conflicts would arise, be resolved and likely arise again, but a pragmatic ethic would constantly evolve and improve.

Like all other theories, pragmatism has its share of critics. The first criticism is that many people have a fundamental problem with the view that there is not some kind of absolute nature to any belief system. People take comfort from the idea that their beliefs are somehow incontrovertibly correct and above question. Many people simply want absolute beliefs to exist, ones that are so fundamental that they are somehow written into the creed of the universe. To that criticism one response is to say that pragmatism may not completely rule out the existence of such truths, however if such absolutes were to be found, the pragmatic method would be no less likely than any of the other theories to find and verify them. Quite simply if a 'best' belief exists, it makes sense that this kind of search would eventually hone in on it.

Another response to pragmatism is that it is simply a model of what people do anyway. People all over the world have beliefs, acquire new information and then accommodate these new ideas to modify their beliefs. To this pragmatists both agree and disagree.[16,17] It is true that people generally use those beliefs that work best for them and modify them to adapt to the real world. However, using a true pragmatic ethic requires a degree of forethought that most people do not use when defining their own belief systems.

Pragmatism requires that changes be made in an intelligent and organized manner, as exemplified by good, scientific inquiry. It is not some ad hoc set of beliefs but is based on the intelligent scrutiny and application of new experiences. This added scrutiny allows pragmatism to avoid stagnation and to constantly evolve to suit its users.

Through the application of a pragmatic ethic, physicians would have a set of beliefs that would be formulated to overcome the major conflicts of Deontologic and Utilitarian theory. They would have discrete guidelines that would be designed to be consistent and practical. They could avoid the conflicts of managing two distinct moral theories and yet have the best tools available for analysing and resolving problems.

6. CONCLUSION

The changing structure of health care in the United States has brought about a need to re-evaluate the ethical theories that physicians use to make difficult decisions. Deontology and Utilitarianism, as evaluated above, have shown that they each have advantages in specific situations, but they also have fundamental problems. Due to the inflexibility of these theories, they are unable to adequately solve these problems within the context of their own superstructure. Physicians are then placed in a position where they do not have a solid moral apparatus in place to make practical decisions on a consistent basis. A pragmatic ethic, by combining a different view about belief systems and an ability to evolve and adapt with experience, is a method that may adequately provide physicians with a structure to make, evaluate and improve their ethical decisionmaking capacities. In addition it has the flexibility to be applied by physicians on a practical level and can be modified to meet their ever-changing needs.

ENDNOTES AND REFERENCES

1. JS Mill, *Utilitarianism* (Indianapolis: Hackett Publishing Company), 1861: 6-8.
2. I. Kant, *Groundwork of the Metaphysics of Morals.* Trans. by HJ Patton (NY: Harper and Row), 1948: 87-9.
3. Ibid., pp. 96-7.
4. EH Loewy *Textbook of Healthcare Ethics* (NY: Plenum Press), 1996: 19-50.
5. Hippocrates, trans.by Francis Adams (NY: Loeb), vol. I, 299-301.
6. American Medical Association, Council on Ethical and Judicial Affairs, "Ethical Issues in Managed Care," *JAMA* 1995 273: 330-5
7. N. Levinsky, "The Doctor's Master," *New England Journal of Medicine* 1984 311: 1573-5

8. J. Paris, "Managed Care, Cost Control and the Common Good," *Cambridge Quarterly of Healthcare Ethics* 2000; 182-188.
9. Hippocrates, trans. by Francis Adams (New York; Loeb) vol. I, 299-301.
10. Idem., J. Paris, 2000.
11. JJ Paris, SH Miles, A. Kohrman, F. Reardon, "Guidelines on Care of Anecaphalic Infants: A Reponse to Baby K," *Journal of Perinatology*,1995; 15:318-324.
12. In the matter of Baby K, 16F.3rd 590 (4th Circuit, 1994)
13. JS Mill, *Utilitarianism* (Indianapolis: Hackett Publishing Company), 1861:6-25
14. KJ Ehlen, G. Sprenger, "Ethics and Decision Making in Healthcare," *Journal of Healthcare Management*, 1998 May-June, 43(3):219-21
15. *The Cambridge Dictionary of Philosophy*, ed. by R. Audi, (Cambridge University Press), 1995:638
16. L. Menand, *Pragmatism: A Reader* (NY: Random House Inc), xii-xix
17. "What Pragmatism Means," *Pragmatism: A Reader,* ed. by W. James, L. Menand (NY: Random House Inc), 93-111.

Chapter 10

The Old Ethics and the New Economics of Health Care

Are They Compatible?

Ben Rich
Associate Professor, Bioethics Program
University of California, Davis
Sacramento, California 95817
e-mail: barich@ucdavis.edu

Key words: health care systems, access to health care, socio-economic considerations, allocation of scarce health care resources, medical ethics, professional accountability, managed care

Abstract: A number of advocates for major reform of the U.S. health care system, most recently and prominently Richard D. Lamm, a former governor of Colorado, have begun to argue vociferously that the prevailing medical ethics that is taught in academic medical centers in this country is antithetical to the reform that must take place if the cost of care is to be reduced and access to care more equitably distributed among the population. In a recent issue of *The Hastings Center Report*, Mr. Lamm, who is a member of the Board of Directors of the Hastings Center, insists that the time is ripe for "redrawing the ethics map." Lamm's chief complaint, which he has repeatedly expressed on the lecture circuit, at conferences, and on commissions, is that medical ethics as it has been traditionally presented, by admonishing the physician to consider always and only the welfare of each individual patient, ignores the realities of budgets and of the limited financial resources available to meet the health care needs of the population.

In this paper I will critique the argument that the prevailing approach to the ethics of medicine is inconsistent with or fails to take account of the cost and resource implications of health care, and dispute the contention that an entirely

Changing Health Care Systems from Ethical, Economic, and Cross Cultural Perspectives, edited by Loewy and Loewy. Kluwer Academic/Plenum Publishers, New York, 2001.

> new ethics of medicine must be created that is sensitive to and informed by the need to impose constraints on procedures and treatments that are very costly and marginally beneficial. I will also maintain that contemporary bioethicists are being unjustifiably held out by Lamm and others as one of the principle causes of or contributors to skyrocketing health care cost inflation on the one hand and gross inequalities in access to health care on the other. Legitimate ethical concerns about the efforts of managed care organizations to constrain costs are being taken out of context and used as examples of how medical ethics is oblivious to the fact and implications of scarce resources. Ultimately, I maintain that there is no need for a new ethical regime in order to confront the very real problems related to the cost of and the access to quality health care in the United States.

To the list of scapegoats for runaway health care costs and the persistent failure of the United States to achieve meaningful reform of its nonsystem of health care delivery has been added (improbably) the influence of health care ethics (bioethics) on clinical decision making and practice patterns. The most recent assault on bioethics of this genre comes in the form of an article in the *Hastings Center Report* by a former Governor of the State of Colorado (1975-87), Richard D. Lamm, who is presently the Director of the Center for Public Policy and Contemporary Issues at the University of Denver and a member of the Board of Directors of the Hastings Center.[1] Governor Lamm has become a stump speaker on health policy of late, and has spoken out from his bully pulpit about what he considers to be the pernicious role of an autonomy-based bioethics. In the article to which I refer, entitled "Redrawing the Ethics Map,"[2] Lamm argues that bioethics has delineated a moral universe in which clinical decision making takes place without regard to economic considerations and fails to provide any useful guide to the formulation of a health policy that is necessarily constrained by limited resources. Consequently, he asserts that the "ethics map" must be "redrawn" so that it comports with the reality of budgets and a proper recognition of the opportunity cost inherent in expenditures on health care, especially those which provide only marginal benefit to the patient.

Any member of this maligned group of bioethicists is likely to experience an almost irresistible impulse to critique Lamm's article on the micro-level, based upon its own terminology and style of argument. If, as Lamm states at the beginning of his article, "medical ethics...is a map to a world," and one which needs to be redrawn, then presumably the need arises because one of two conditions obtains. Either the world that the map represents has changed, such that the map no longer accurately reflects reality, or that the world has not changed but the map was erroneous *ab initio*, and should be redrawn so that it will then comport with the world as it has always been. The former situation is analogous to a map of the world from its earliest origins, before the continental drift. The latter analogous to a map of the world from the era in which it was still believed to be flat. While Lamm suggests it is the former, one cannot be absolutely certain. I fear that the map analogy is fundamentally flawed, and will not bear the weight of close analysis. Unfortunately, Lamm does not point to any particular bioethical principles or practices as prime examples of the problems to which he more generally alludes. However, I suspect that respect for patient autonomy is probably number 1 on his hit list. In other settings (Lamm and this author, until recently, both served on the Colorado Governor's Commission on Life and the Law) he has taken bioethicists to task for browbeating physicians into the mindset that respect for patient autonomy uniformly demands that patient's receive everything they demand, without regard to any other considerations, and that it is in fact unethical for a physician to factor cost, especially opportunity cost, into the clinical decision making process. To the extent that some practitioners do possess such a mindset, I suggest it is born of a fear of potential malpractice liability rather than intimidation by bioethicists.

Lamm states: "I cannot express my frustration at sitting in a hospital ethics meeting, agonizing over whether to recognize a living will and knowing that within blocks there are medically indigent citizens with very restricted access to *any* health care."[3] Of course, since Lamm is also an attorney, he cannot be suggesting that a patient's living will, under current law, should have no influence over the medical treatment provided. After all, the Colorado advance directive statute provides that the failure to follow a valid advance directive or to transfer the care of the patient to another physician who will may be deemed unprofessional conduct and subject the physician to disciplinary action by the state medical licensing board.[4] Moreover, most advance directives seek to anticipatorily decline marginally beneficial treatment, not demand futile interventions. Nor, I hope, is he suggesting that the bioethicists at that meeting should have declined an appearance and instead hurried on down to the legislature to lobby for increased funding of the medically indigent program. Some of us, as the colloquialism characterizes it, have "been there/done that," and have been confronted with the magisterial indifference of fiscally conservative legislators who are fundamentally opposed to such governmental programs.

In his article, Lamm also decries the fact that Karen Ann Quinlan was maintained on life support in a state (New Jersey) that leaves 14 percent of its citizens without basic health care. Of course, what he does not mention is that in such cases, and particularly in the case of Nancy Cruzan, which her family pursued all the way to the U.S. Supreme Court,[5] it was the state that was asserting an unqualified interest in preserving human life, one which was alleged to be sufficiently strong to override the Cruzan family's judgment that Nancy would not wish to be kept alive in a persistent vegetative state. Indeed, the State of Missouri in *Cruzan* insisted that because Nancy was being cared for in a state facility at state expense, cost was not a consideration and hence should not be an issue in the case. Recall that Nancy

Cruzan, by the time her case reached the Supreme Court, had been in a persistent vegetative state for nearly 7 years. The State of Missouri did not seem to be arguing that continuing artificial nutrition and hydration for the additional 40 years of her life expectancy would confer a benefit upon her, but rather that it would provide a powerful symbolic expression of the state's ethical stance that all human life is sacred regardless of its quality.[6] So much for the purported legislative responsibility for careful stewardship of scarce public resources.

The vast majority of bioethicists strongly supported the Cruzan family's efforts to exercise surrogate decision making in a manner they believed to be consistent with her wishes and values. It was the public policy of Missouri, heedless of both the cost/benefit implications of continued treatment and of the prevailing principles of medical ethics that resulted in the provision of years of costly, unwanted, and nonbeneficial care. This point must be strongly emphasized because in his article Lamm would have the reader believe that but for the unrealistic admonitions of bioethicists, those who formulate and implement public policy would have reformed health care delivery in a more ethical, just, and fiscally responsible fashion. He quotes with approval one Alan Williams: "anyone who says that no account should be paid to cost is really saying that no account should be paid to the sacrifices imposed on others."[7] Yet as we see in the Cruzan case, and many others like it in other jurisdictions, it was the public policy makers of the state who, in their missionary zeal to establish the sanctity of life principle as an article of faith, took no account whatsoever of those citizens of the state who were desperately seeking but could not obtain basic health care, those whose welfare Lamm is purportedly obsessing over during his foray into hospital ethics rounds.

I fear that further pursuit of the host of *non-sequiturs*, false dichotomies, and straw man arguments that permeate this short piece will lead us down many blind alleys. Instead, I propose to take Lamm as both serious and sincere when he posits "the state's duty to all the medically indigent," which to me reasonably implies a social responsibility to allocate scarce public resources in such a fashion that the indigent are provided with reasonable access to some minimal level of health care and its moral implications. Now I would be among the first to assert that American bioethicists generally, as well as most health care professionals, have devoted woefully inadequate time, attention and energy to the travesty ("problem" is much too mild a term) of the millions of Americans without access to care. Unquestionably, the vast majority of bioethical discourse in the United States is focused on the many negative rights that have been asserted *in* healthcare, and not on any single, preeminent, and foundational right *to* healthcare. Nevertheless, prevailing principles of bioethics have neither caused nor perpetuated this situation. The ethical ground of such a duty on the part of the state is, at least implicitly, one of the four pillars of the "Georgetown mantra" of principles, to wit, justice. Indeed, almost twenty years ago the President's Commission for the Study of Ethical Problems in Medicine and Biomedical and Behavioral Research issued a report which concluded that: "society has a moral obligation to ensure that everyone has access to adequate care without being subject to excessive burdens."[8] It should not have escaped Governor Lamm's notice that the professional staff of the President's Commission, including its Executive Director, was comprised of some of the most prominent bioethicists in America today.[9]

Lamm seems to think that he has found at least one kindred spirit in academic bioethics, Haavi Morreim, and he is quite fond of quoting from her book *Balancing Act: The New Medical Ethics of Medicine's New Economics*.[10] It is true that she seeks to provide ethical legitimacy for the increasing involvement in medical decisionmaking of the "economic agents" representing third party payers, as well as to include among the physician's professional responsibilities consideration of the limits of health plan coverage and the competing interests of other patients.[11] But as part of the "balancing act," she insists that if patients are to be made increasingly responsible for maintaining their own health and for making wise choices in securing health care, then they must have real options from which to choose and the information necessary to select from among them. In a deftly noncommittal statement she adds: "Quite likely we would want to see such choices take place within a system that assured universal access for all citizens."[12] The reason, though she leaves it unstated, is presumably that those without any adequate discretionary income can have no genuine options, only the false and unacceptable one of choosing which to forego—food, shelter, or health care.

Ultimately, the title of her book to the contrary notwithstanding, Morreim does not really claim that the new medical economics demands a radically new and different medical ethics, one which fundamentally transforms the normative dimensions of the physician-patient relationship. Acknowledging that patients with maladies requiring medical interventions in the new economics of managed care are at least as, if not more vulnerable than similar patients in the former fee-for-service system, she maintains that the patient needs a "powerful, knowledgeable friend who can help him to understand his choices and the impact they have on his life, and who can help him to secure the care to which he is entitled. The physician therefore remains a fiduciary with powerful, even if not limitless, duties to do his best for his patients."[13] The "revised ethic," she concludes, "probably more than anything else must emphasize communication [between physician and patient]." In a description that appears to characterize the physician-patient encounters as they presently take place in many managed care settings, she warns:

> If the physician helps the patient to find his way through such choices, he needs to engage, not in perfunctory recitations of fact or lengthy iterations of listed options, but in careful conversation—a conversation that will require the patient to contribute actively, and not stand passively by. Because patients do remain vulnerable and in many cases very intimidated by the medical environment, the physician is also obligated to make such discussions as easy and inviting as possible. A perfunctory "are there any questions?" asked with one hand on the door knob and one foot out the door, will not suffice. The physician must strive to elicit the patient's real concerns—the ones he may be too embarrassed to raise or too confused yet to have identified—in order to engage him in the kind of vigorous dialogue that is essential if all the important benefits and burdens of care are to be realized.[14]

I have quoted Professor Morreim at some length in order to make perfectly clear that the "new medical ethics" which she suggests is required by "medicine's new economics" is not one which turns upside down the prevailing principles of medical ethics as they have developed over the last 30 years in the United States. Indeed, all of this meaningful and informative ("careful") conversation, which she deems essential to the "new medical ethics" appears to be little more than the faithful adherence to the shared decision making model of the physician-patient relationship, the model necessitated by Lamm's public enemy number one—the bioethical principle of respect for patient autonomy, a condition precedent to the exercise of which is full disclosure of all information relevant to the condition and options for treating it.

We come, then, to the issue that has only been mentioned in passing by Governor Lamm and Professor Morreim, but which in my judgement should be the cornerstone of any American Bioethics for the New Millennium, i.e., universal access. The pursuit of cost-savings and the elimination of nonbeneficial or marginally beneficial health care are not unqualifiedly moral objectives in the context of contemporary American health care, especially if the patient is enrolled in a for-profit HMO. The dollars that the organization saves by compelling physicians to see a patient every 5-8 minutes or by eliminating new and expensive prescription medications from its formulary do not necessarily benefit the plan enrollees in some other fashion. Too often they are funneled into exorbitant executive compensation packages, stockholder dividends, or ridiculously high administrative costs. For example, in 1996 the aggregate income in salary and stock options of the eight highest pain CEO's of managed care organizations in the United States was nearly $1 billion, or enough to provide healthcare for one year to 2.7 million Americans.[15] The average overhead cost for these same organizations was an exorbitant 21%, compared to 2% for Medicare from 4-5% for Medicaid, which have been repeatedly labeled "cumbersome and bureaucratic" by the private insurance industry.[16] Furthermore, no set of patient behaviors is more "irrational" in a strict sense, yet at the same time understandable, than those to which the uninsured in the United States are forced to engage, i.e., going without low cost preventive care, only to later present to the emergency rooms of so-called "safety net" hospitals when they are *in extremis* and requiring expensive rescue medical measures with little likelihood of ultimate success.

In one sense Governor Lamm is correct when he charges contemporary American Bioethics with an unconscionable obsession with the individual patient and a concomitant disregard for the larger community of which that patient is a part. He insists that the costs to that larger community of the treatment decisions that are made by or on behalf of each patient should legitimately be factored into the process. With that I do not totally disagree. By the same token, however, that community, in this country, bears moral responsibility for the fact that there are 45 million people without access to basic healthcare. Curiously, while he is railing at the moral myopia of American bioethicists, Governor Lamm fails to bring the problem of the uninsured into focus and give it the attention it deserves. I earlier alluded to the report of the President's Commission *Securing Access to Health Care* and the qualified support which it offered for universal access. Yet it has been noted that the Commission, despite the guidance and counsel of a cadre of distinguished bioethicists, was unable to agree upon an ethical framework for universal access to health care.[17] Similarly and unsurprisingly, given their much more marginalized role, the "Ethics Cluster Group" of President Clinton's ill-fated Health Care Task Force a decade later was unable to articulate an ethical basis for universal coverage.[18]

George Annas observes: "over the last 30 years of bioethics almost no one in the field has discussed what is perhaps the central ethical flaw in our health care system: the lack of insurance and access to care of 40 million Americans."[19] The silence on this issue, which should be seen as having transcendent ethical and public policy implications, is exemplified by but certainly not limited to bioethicists. It permeates all public discourse. For example, in the summer of 1999, Congress and the Clinton administration debated what to do with the trillion dollar plus federal budget surplus that is anticipated in the first decade of the new millennium, as well as what form a federal Patient's Bill of Rights should take. Both Republicans and Democrats conceded that some of the projected surplus should be earmarked for the purpose of "shoring up" the Medicare Trust Fund. Thus, the entire focus of the health care reform policy of both parties was on securing health care benefits for those who already have them, and recognizing negative rights *in* healthcare that can be successfully asserted against managed care organizations. No major figure in any political party—other than Democratic presidential candidate Bill Bradley—or in any responsible governmental position, has had the insight or the temerity to argue that as a matter of ethics and humane public policy some portion of this massive surplus should be used, to borrow the phraseology of the President's Commission, to fulfill this society's "moral obligation to ensure that everyone has access to adequate care."

Nor is there any moral outrage on the part of the citizenry of this country, taking public policymakers to task for their failure to proffer and promote plausible solutions to this national disgrace. One might be tempted to attribute this public indifference to the plight of those without access to health care to some uniquely American perception that this group is made up of people on the fringes of our culture or who are simply unwilling to get a job and earn a living that would enable them to purchase health insurance. But well over half of the 45 million are employed, and many of those at more than one job.[20] As a few have noted, they are probably some of the hardest working among us: single parents, service-sector non-unionized employees, and small business workers. These are people we encounter and interact with every day, and with whom it should not require a super-human effort to identify and empathize.

One way in which to think about American democracy is as essentially a kingdom of rights, its *Magna Carta* being the first ten amendments to the Constitution. Significantly, for purposes of the issue before us, they are often framed as the negative rights of the citizen as against the state, delineating those things which a free society and its governmental institutions cannot do *to* them. The most fundamental of these rights is, in the words of Justice Brandeis: *the right to be let alone*—the most comprehensive of rights and the right most valued by civilized men."[21] For the very young, the very old, the disabled, and the impoverished, there is a dark side to the Bill of Rights. The right to be let alone becomes the right to be neglected and discounted. Our collective obsession with this right has undermined any sense of social obligation toward fellow citizens who are disadvantaged by nature or circumstance.

Communitarian political philosophy has extensively critiqued this atomistic view of the nature of persons and how they live their lives. But in the United States today, if one is among the 45 million who can neither afford health insurance nor qualify for Medicare or Medicaid, then "atomistic" is a term that accurately describes the nature of their circumstance. The President's Commission characterized the United States as "a society concerned not only with fairness and equality of opportunity but also with the redemptive powers of science [in which] there is a *felt obligation* to ensure that some level of health services is available to all."[22] Yet, as just noted, at the very moment when this nation contemplates the disposition of a budget surplus of unprecedented and gargantuan proportions, we hear no call within the corridors of power for any health care reform to eliminate the uninsured.

Like Diogenes in pursuit of an honest man, we search in vain for any real evidence of the "felt obligation" of this society to insure basic health care for all citizens.

The Communitarian social ethic is alive and well in all of the other industrialized nations. There has been no need for a new cadre of professionals—bioethicists—to foster or advocate for such a "felt obligation" to provide minimal health care for all members of society. And in the United States, which has purportedly been "blessed" with such an elite cadre of professional ethicists for the last 25 years, as Professor Annas points out, the ethical implications of millions of people without access to care is not even an issue for discussion, let alone a priority item on the agenda of national bioethics conferences. Those agendas focus, with compulsive regularity, on the hot button, megawatt issues that earn bioethicists a 30-second soundbite on the local news. Obsessing on such matters at times puts bioethicists in the bizarre position of talking seriously about the right of all terminally ill persons in our society to *die* with dignity without ever having possessed the insight or temerity to posit the right of all persons in that same society to *live* with dignity. At the very least, the latter requires universal access not only to some minimally sufficient level of health care and education, but also to the other necessities of living as well, such as adequate nutrition, shelter, and employment that provides a living wage. Remarkably, in the most affluent nation that has ever existed on this planet, 15% of men and 25% of women working full time have incomes that are below the poverty level.[23]

Organized medicine has, historically, been part of the continuing problem of access to care when one might have hoped that the collective social responsibility of the physician would demand unstinting advocacy for universal access to care. In the words of Larry Churchill: "As a group, physicians have lobbied intensively for the current structure of health care" and hence are responsible "not only for the quality of care they give...but also for the characteristics of the access system within which they work."[24] Each time a proposal to move toward universal access to health care has been made, usually involving the creation of a single payor system, but always with some increased involvement of government, organized medicine has lobbied legislators heavily and effectively in opposition to it.[25] To cite only one example, in 1944 President Roosevelt asked Congress to affirm an "economic bill of rights" which would include a right to adequate medical care. Shortly after becoming President, Truman repeated the request and called upon Congress to pass a national program that would assure the right to adequate medical care. The response of the American Medical Association (AMA) in an editorial was to declare that Truman's national health insurance plan would make doctors "slaves." When Truman shortly thereafter won his surprise victory in the presidential election, the AMA launched what was at the time the most expensive public relations campaign in history to defeat "socialized medicine."[26]

Interestingly, in their 1981 book *A Philosophical Basis of Medical Practice*, Pellegrino and Thomasma maintain that an "unequivocal dedication to a system of care, geographically and temporally accessible at all times, and designed to respond to felt needs for medical assistance...[is] an obligation incurred by all of us by our mutual declaration of the kind of society we profess to be...."[27] They go on to state:

> We have, in a sense, all made a set of mutual promises to guarantee to each other a certain kind of society, one which is sensitive to and secures those things closest to our needs as humans. We would break our communal promise, tell a communal lie, and live an inauthentic social life if we neglected to exert every effort to assure the minimum security of access to primary care whenever it is needed.[28]

While these fine words sound very much like a moral manifesto for the profession of medicine to engage collectively in the diligent pursuit of universal access to basic healthcare, the author's waffle on the brink of such a commitment. They pose the following rhetorical questions: "Is there a corporate [medical] responsibility to make these "professions" authentic? Are not all members of the health professions to some extent culpable if the aggregate of their efforts neglects a fundamental need?" While a strongly affirmative answer would seem to be required by the previously quoted passages, Pellegrino and Thomasma lamely declare: "These are legitimate ethical questions and the substance for an ethical debate just beginning to become public."[29] Yet we know from what has already been discussed that such a debate has never taken place, and the question continues to be begged as to the professional responsibility of physicians and bioethicists to continually demand it.

At this time there is no need to become bogged down by difficult technical questions such as what constitutes an adequate level of care to which all should be entitled, what such a minimal level of care would cost, or how it might be most effectively and efficiently provided. But the experience of other Western industrialized nations with universal access controverts the proposition that the recognition of such a positive right or social obligation necessarily places us on a slope too steep to set workable limits. Indeed, reliable studies substantiate the feasibility and affordability, indeed the net cost savings of such plans. For example, the Congressional Budget Office estimates that financing a single payer national health plan with modest co-pays would require a tax increase of $856 per capita, but that it would also decrease private sector spending by $910, resulting in a per capita savings of $54. Arguments against the feasibility of a universal system of healthcare are specious, and at least in part designed to use logistics and economics as the clubs with which to beat back the fundamental ethical issue—whether American society believes, as the President's Commission presumed, that "the special nature of health care dictates that everyone have access to *some* level of care: enough care to achieve sufficient welfare, opportunity, information, and evidence of interpersonal concern to facilitate a reasonably full and satisfying life."[30] By "special nature" the Commission means the existence of certain properties or characteristics of health care that make it particularly significant both in the life of the individual and the life of the society. It is the possession of those special properties which, upon careful examination, lead inexorably to "the ethical conclusion that it ought to be distributed equitably."

The Commission enumerates and elaborates upon four properties of health care that endow it with a special status: well being, opportunity, information, and the interpersonal significance of illness, birth and death. Personal well being is significantly related to the state of one's mental and physical health, and so too is equality of opportunity, a value that since the civil rights movement has been touted as a hallmark of American society. Information, as the previously quoted statement of Professor Morreim reveals, is essential to the ability of persons to make responsible decisions about their health. Imparting accurate and complete information to patients is a fundamental responsibility of all health care professionals. Finally, and perhaps most significantly for purposes of this paper, the Commission maintains that:

> health care has a special interpersonal significance: it expresses and nurtures bonds of empathy and compassion. The depth of a society's concern about health care can be seen as a measure of its sense of solidarity in the face of suffering and death ... For these reasons a society's commitment to health care reflects some of its most basic attitudes about what it is to be a member of the human community.[31]

We must be careful in how we measure "the depth of a society's concern about health care." One could put forward the argument that because in 1990 the United States was responsible for 41% of the world's total health care expenditures while at the same time having less than 5% of the world's population, that is conclusive evidence of our concern about health care. But such an assertion becomes ludicrous as "a measure of its sense of solidarity in the face of suffering and death," given the corresponding fact that as of that date there were between 30 and 40 million Americans who received the benefits of little or none of those health care expenditures. From an ethical standpoint, how ought we to interpret those statistics in light of the fact that not only do we spend more on health care than any other nation, but that we spend more on health care than on anything else? A logical yet indefensible conclusion is that we believe that health care is of overriding importance to us as individuals, but we have no sense whatsoever of participation in a human community in which we have a shared responsibility to insure that something of such overriding importance is available to all.

We are the wealthiest nation on earth, with the most physicians per capita and the most sophisticated medical technology. There are simply no financial, technical, or other practical reasons why the United States of America continues to allow millions of citizens to go without health care while other nations with significantly less wealth mange to provide universal coverage without sacrificing any of the other major responsibilities of government. The reasons are political, and by implication moral. They say nothing about what we have in a material sense, but they say everything about who we are as a people in a civic, moral and spiritual sense. "We have," to quote the cartoon character Pogo, "seen the enemy and he is us."

At the conclusion of his article, Governor Lamm insists that "[t]he view I am affirming is not as harsh as patient advocates might think. It reflects, not an absence of caring, but a broader definition of caring. Public policy is compassionate, not an individual at time, but in view of total unmet social needs."[32] One of the primary unmet social need in the United States is today and has been for decades the millions without access to health care. Public policy has not only tolerated this deplorable situation, but has also persistently refused to take definitive action to end it. Medicine and bioethics have been complicit in this total abdication of moral and civic responsibility. The tension Governor Lamm posits between an altruistic group of public policy makers seeking to address the problem of those without access to care and a hidebound cadre of bioethicists insisting that health care costs cannot be managed and care cannot be rationed without violating fundamental principles of medical ethics is not only bogus, it is utterly ridiculous. Public policy makers, organized medicine, and bioethicists have been co-conspirators, or at least fellow travelers, in the perpetuation, through a not-so-benign neglect, of a national disgrace. But it has nothing to do with some flawed map of the domain of medical ethics. The flaw, and it may well turn out to be a fatal one, is in this nation's impoverished conception of social responsibility. In making their case for a social obligation to provide everyone with access to basic care, Pellegrino and Thomasma explicitly disclaim that they need to invoke any new ethical principles, or to paraphrase Governor Lamm, that they are redrawing the ethics map in order to justify a social obligation to provide access to basic healthcare. In their view, and it is one I share, the promises inherent in our formative social and political structures, the special vulnerability and the universal fact of illness, and the traditional ethical principles of promise keeping, truth telling and justice are sufficient.

ENDNOTES

1. Governor Lamm came to national attention with regard to health policy years ago when he was quoted (or misquoted, as the case may be) that elderly have a "duty to die," i.e., to decline heroic life-extending measures which offer marginal benefit and divert scarce health care resources from more beneficial use, such as pediatric care.
2. RD Lamm, "Redrawing the Ethics Map," *Hastings Center Report* 1999; 29: 28-29.
3. Ibid., p. 28.
4. CRS 15-18-113 (5) (1987).
5. *Cruzan v. Director, Missouri Dept. of Health*, 497 U.S. 261 (1990).
6. The Missouri Supreme Court opinion in the *Cruzan* case was perhaps more revealing than its authors intended. The court seemed to recognize a need to justify the fact that it was forcing Nancy Cruzan (and vicariously her family) to endure the indefinite prolongation of her persistent vegetative state as an example of the sanctity of life principle which the State of Missouri had adopted. Hence it stated:

    Given the fact that Nancy is alive [physiologically] and that the burdens of her treatment are not excessive for her [inasmuch as she is permanently unconscious], we do not believe that her right to refuse treatment, whether that right proceeds from a constitutional right of privacy or a common law right to refuse treatment, outweighs the immense, clear fact of life in which the state maintains a vital [one might say vitalist] interest. *Cruzan v. Harmon*, 760 SW 2d 408, 424 (1988).
7. Alan Williams, "Cost-Effectiveness Analysis: Is It Ethical?" *Journal of Medical Ethics* 1992; 18: 7-11, p. 7.
8. President's Commission for the Study of Ethical Problems in Medicine and Biomedical and Behavioral Research. *Securing Access to Health Care.* Washington, D.C. 1983: 22. While it is true that the Commission urged that "society" be construed broadly to include not only the state but "the collective American community," it also concluded that "the Federal government has the ultimate responsibility for seeing that health care is available to all...." *Ibid.* p. 5.
9. In addition to Alexander M. Capron, the Executive Director, the professional staff included Assistant Directors Joanne Lynn and Alan Meisel, as well as Dan Brock, Allen Buchanan, and Daniel Wikler.
10. E. Haavi Morreim, *Balancing Act: The New Medical Ethics of Medicine's New Economics* (Dordrecht: Kluwer Academic Publishers), 1991.
11. Ibid., p. 2.
12. *Ibid.*, p. 145.
13. *Ibid.*, p. 148.
14. *Ibid.*, pp. 149-150.
15. *Managed Healthcare Market Report*, 1/31/97.
16. Corporate Research Group, *Outlook for Managed Care*, 1997.
17. GJ Annas, "The Dominance of American Law (and Market Values) over American Bioethics." *Meta Medical Ethics: The Philosophical Foundations of Bioethics*, ed. by MA Grodin (Dordrecht: Kluwer Academic Publishers), 1995: 92.
18. N. Dubler, "Working on the Clinton Administration's Health Care Task Force." *Kennedy Institute of Ethics Journal*, 1993; 3:421-431.
19. Note 7, *op. cit.*
20. L. Gage, L. Fagnani, J. Tolbert, CC Burch, *The Safety Net in Transition: America's Uninsured and Underinsured: Who Cares?* (NY: National Association of Public Hospitals and Health Systems), 1998, p. 12.

21. *Olmstead v. United States*, 277 U.S. 438, 478 (1928) (Brandeis J., dissenting).
22. Note 4, *op. cit.*, p. 12.
23. Census Bureau. Workers with Low Earnings: 1964-1990 and Tabulation from 1997 CPS.
24. L. Churchill, *Rationing Health Care in America: Perceptions and Principles of Justice* (Notre Dame, IN: University of Notre Dame Press), 1982, 108-109.
25. See, e.g., P. Starr, *The Social Transformation of American Medicine*. (NY: Basic Books), 1982, Book 2, Ch.1.
26. *Ibid.* at pp. 284-286.
27. E. Pellegrino, D. Thomasma, *A Philosophical Basis of Medical Practice*, (NY: Oxford University Press), 1981: 235, 239.
28. *Ibid.,* p. 243.
29. *Ibid.*, p. 241.
30. Note 8, *op. cit.*, p. 20.
31. *Ibid.*, p. 17.
32. Note 2, *op. cit.*, p. 29.

Chapter 11

Playing the HMO Language Game

Repatriation, Extradition or Deportation?

Roberta Springer Loewy† and Erich H. Loewy‡
†Assistant Clinical Professor, Bioethics Program
University of California, Davis
Sacramento, California 95817
e-mail: roberta.loewy@ucdmc.ucdavis.edu

‡Professor and Endowed Alumni Association Chair of Bioethics
Associate Professor, Philosophy
University of California, Davis
Sacramento, California 95817
e-mail: ehloewy@ucdavis.edu

Key words: health care systems, euphemism, managed care, patient-clinician relationship, professional education, fiscal responsibility, socio-economic determinants of practice, patient best interests, conflicting values/obligations

Abstract: In emergency situations, patients who belong to one managed care organization are not rarely admitted to hospitals outside of those "recognized" by the organization to which the patient has subscribed or been assigned. In such situations every attempt is made by the managed care organization to which the patient is said to "belong" to effect a transfer from the admitting hospital to the particular "preferred provider" facility (PPO) of the health care organization or managed care plan. The managed care system calls this procedure "repatriation." Pressure is brought to bear on the admitting hospital to transfer such patients at the first possible moment—sometimes with life sustaining equipment in place—and always in the middle of treatment. Such transfers are often against patients' wishes. More importantly, because of the disruption in the continuity of professional caregivers they necessitate such transfers are also often against patients' best interests. These disruptions are especially troubling when they occur within teaching hospitals where, in addition to be-

Changing Health Care Systems from Ethical, Economic, and Cross Cultural Perspectives, edited by Loewy and Loewy. Kluwer Academic/Plenum Publishers, New York, 2001.

ing burdensome for patients, they hamper progress in research and in teaching and learning for both students and faculty alike.

The authors suggest that the adoption of the term, "repatriation," is both linguistically inappropriate and ethically problematic. It is linguistically inappropriate because what occurs under such "repatriation" more closely resembles what actually occurs under "extradition"—or even "deportation." And it is ethically problematic because the language of "repatriation" is invoked to suggest a community of caring concern for patients to hide what, in fact, are little more than bald cost-cutting measures. Such measures are clearly intended, first and foremost, to benefit the particular health care organization—and their shareholders—often at the patient's expense. The authors argue that such measures not only disrupt the continuity of care of individual patients but also threaten, as well, the integrity of both the profession of medicine and the community itself.

CONTEXT

Prior to our introduction, we find it timely to offer several remarks about the particular social arrangement of access to and delivery of health care that currently exists here in the United States. First, we stand alone among modern democratic nations[1] in our denial of equitable access—to some degree, however small—to basic health care services for all *persons* (N.B.: all persons, not just all citizens) living within our borders. Second, we have no health care "system" within this country, but only the most inequitable, inefficient and costly hodge-podge of public and private arrangements imaginable. Here are just a few shameful examples gleaned from some of the latest government statistics:

- In 1996, 14.0% of children under age 18 were completely without health insurance
- In 1997, 16.8% of persons under age 65 (41 million persons) were without health insurance
- In 1997, only 69.6% of persons age 65 and over (22.3 million) had both Medicare and private insurance
- In 1997, 20.7% of persons age 65 and over had only Medicare
- In 1998. 76.6 million persons were enrolled in Health Maintenance Organizations [2]

While these few statistics are bleak enough, they fail to reflect the true nature of "access" to health insurance. Many of us who do have health insurance remain woefully underinsured or are insured only for the more catastrophic types of health care interventions, with little (if any) coverage for preventive care or early medical intervention. The phrase "having health insurance" can refer to having comprehensive coverage (increasingly unlikely today) to having minimal coverage (more likely) to having spotty, useless or duplicate coverage, any of which may be—and often is—grossly inappropriate to the particular needs of the person being insured. In addition, those of us who are insured have increasingly less influence over the conditions under which we are insured. With these depressing data freshly in mind, we now begin our discussion of the language—and the realities—of "managed care."

INTRODUCTION

More than any other innovation in professional medicine in the United States, the recent phenomenon of "managed care" (even here, language is manipulated to suggest that care was unmanaged before!) poses a serious threat to the traditional notion of the patient/physician relationship. Of course, the direct, relatively uncomplicated one-to-one relationship between physician and patient is, in reality, an ideal social arrangement away from which the actual practice of medicine has been evolving for decades. Two primary factors that have contributed to this evolution are (1) scientific advances and (2) efforts to provide indemnity against the bio/psycho/social burdens associated with disease, illness and disability (i.e., various health care insurance and entitlement schemes).[3]

The first factor (scientific advances) has certainly widened the gap between patients and physicians by necessitating increasingly more complicated, technologically sophisticated and labor-intensive treatment—to say nothing about the increase in education and research necessary to support such practices. As a result, a variety of subspecialties in the various fields of medicine, nursing and allied health care has arisen. Not unexpectedly, this has led to problems in communication—not only between patient and physician, but between the various members of the health care team as well. Such problems can seriously compromise the development of an optimal plan of care for a patient.

And yet, these scientific changes have not unduly disturbed the basic sets of mutual and reciprocal obligations traditionally recognized to exist between patients and physicians: obligations that have always included such notions as trust, fidelity, beneficence, non-maleficence, respect, however variously interpreted.[4] In other words, while the first factor complicated the one-to-one relationship between patient and physician, the nature of the healing relationship still remained essentially dyadic: professionals committed to the provision of medical care on the one hand and the beneficiaries of that care on the other.

Thus, irrespective of the actual numbers of individuals involved, this first factor has not created insurmountable difficulties for the healing relationship since its participants still share the same general interests and commitments regarding its goals. And high on that list is the development of practices and expertise conducive to the patient's bio-medical good. Most importantly, the means by which this good is brought about both sustain and are sustained by the basic sets of mutual and reciprocal obligations that characterize the relationship. As a result, what differences arise, arise—by and large—within the intimacy of the healing relationship whose antecedent conditions and consequences are overwhelmingly governed by a single social institution: medicine.

The second factor (efforts to provide indemnity against the bio/psycho/social burdens associated with disease, illness and disability) has also inadvertently widened the gap between patients and physicians—though in a much different way. It has done so by importing into the dyadic patient/healer relationship (which already includes an extensive professional health care team) bureaucratic representatives of other social institutions whose interests and goals are necessarily larger than—and, at times, may even directly conflict with—the narrower interests and goals of the primary, dyadic relationship between patients and healers. These bureaucratic representatives are not—nor can they be—bound by the same basic sets of mutual and reciprocal obligations that bind patients and healers. Rather, they are bound by obligations external to the healing relationship, i.e., obligations that are specific to the particular social institution they represent.

Thus, like the first, this second factor has certainly widened the gap between patients and their caregivers. However, unlike the first, the inclusion of these bureaucratic representatives can constitute a threat to the dyadic healing relationship and the mutual obligations it has, for so long, embodied. That is, these additional roles not only alter but, depending on how they are instituted and employed, can actually undermine the basic sets of mutual and reciprocal obligations that have come to characterize and sustain not only specific healing relationships, but the very institution of medicine itself.

The first of these new roles—the health care insurer—has, in one form or another, been in existence for quite some time.[5] As early as 1883 a national, compulsory "sickness insurance" to cover all workers was instituted in Germany by Bismarck, an otherwise rather staunch conservative. For Bismarck, such a guarantee to all members of the community that certain basic needs would be met offered the greatest likelihood of preserving and enhancing social solidarity. The rationale behind this move—that both individuals and their communities flourish when social institutions are structured in such a way that no individual's basic needs go unmet—continues to be the overwhelming choice of most democratic communities today.

In the United States, however, the justification for indemnity against disability and illness has consistently relied less on the idea of social solidarity (and the notion of public, communal responsibility it entails) and more on the idea of rugged individualism (and the notion of private and personal responsibility for oneself and one's immediate family that it entails). As a result, health care insurance in the United States has developed as a hodgepodge of private and public, for profit and not for profit consumer and producer cooperatives, all interested—though for quite different economic and social reasons—in providing various degrees of coverage for health care services.

This fundamental difference in perspective between most other democracies and the United States helps to explain why the second of these new roles—the health care broker—is still so uniquely an American phenomenon. Health maintenance organizations (HMO's) are prime examples of health care brokers because, with their introduction, important distinctions between health care insurer, health care provider and those insured have been effectively blurred. Essentially, HMO's underwrite or "insure" a service. However, the terms of this service are not brokered directly between the HMO and those individuals ultimately being insured, but between the HMO and the *employers* of those being insured. That is, the purchaser is not the person being insured and there is little reason to assume that the motivations, interests or goals of the purchaser are co-extensive with those *being* insured. As a matter of fact, under managed care today both the purchasers of health care insurance and health care providers are much more likely to have more interests in common with the insurer/broker than they do with those being insured.

By employing their own teams of health care providers—thus creating a direct and primary employer/employee relationship with the health care team itself—HMOs are in a quite unique position. They now have more control over the structure and content of health care delivery (and thus, the patient/healer relationship) than either patients or healers. This situation has seriously undercut the ability of either healers or patients to represent their own particular interests adequately, much less the interests of either individual patient/healer relationships or the institution of medicine as a whole. Apace with this change has come a striking change in language: health care professionals are now "providers" and patients are now health care "consumers" (though to corporate insiders we are also known as "covered lives," or even worse, "units of care"!). In short, everything from the physical structure of the patient/healer encounter to the quality of the information exchanged within those encounters to the very structure of the institution of medicine itself has, along with the language we use to describe it, been affected—at times, profoundly so.

With the rise of health maintenance organizations any direct fiduciary relationship between insurer and those actually insured is rapidly disappearing. In the brave new world of health care insurance, the older and direct fiduciary relationship between insurer and those insured has been replaced by an independent brokerage system between insurers and purchasers of health care. This system is indirect in the sense that the fiduciary relationship for insurance is now between insurer and purchaser. It is independent in the sense that, for the insured, the content of the fiduciary relationship is largely a *fait accompli*: its terms have been structured by prior agreement between purchaser and insurer with little, if any representation of those actually being insured. Thus those insured lack, in crucial respects, equal standing in the fiduciary relationship with the insurer and the purchaser.

Such changes threaten to undermine not only the healing relationship between individual patients and their individual caregivers, but the tacitly sanctioned tradition of teaching, research and practice (i.e., patient care) upon which the very art and science of medicine, as a professional institution, is based.[6] In short, we will argue that "repatriation" is one of the more flagrant examples of how the narrower, entrepreneurial interests and goals of our current managed care organizations can thwart what patients, the public and health care professionals have traditionally identified and tacitly endorsed as the legitimate interests and goals of medicine. We turn, now, to an examination of the language of "repatriation."

THE DELIBERATE MISUSE OF LANGUAGE

Managed care's attraction to the analogy of repatriation—the return to the country of one's birth or citizenship—is, perhaps, understandable. The use of the term evokes a number of warm and fuzzy feelings about the benefits of membership in a group—and, obviously, managed care organizations would prefer to consider membership in their group a beneficial thing! However, the usual presumption behind repatriation is that the person who initiates it considers it a desired—and desirable—goal. That is, while everyone comes from some country, repatriation to that country is a voluntary undertaking: the person who actively seeks to be repatriated is the person who initiates it. However, the reality of repatriation under managed care is that patients rarely consider it desirable and even more rarely voluntarily initiate it.

"Repatriation" that is not voluntarily initiated by the subject is not, in fact, repatriation. If anything, it would be more like what occurs in extradition, i.e., the legal surrender of a subject by one state, country or government to the jurisdiction of another state, country or government. This is the case because, in contrast to repatriation, the subject being extradited is passive, the *object*, if you will, of such an action. In most instances, the person being extradited neither desires, nor finds desirable, the thought of being "repatriated." It would be safe to assume that managed care organizations would prefer not to have their "members" feel as though they are unwilling subjects! And while membership obviously entails a sense of "belonging," it certainly is not "belonging" in the sense of "being owned" but, rather, in the sense of "being a respected member of" a group. What has been allowed to occur today under managed care looks increasingly more like what occurred under feudalism in the middle ages! Patients are much like serfs tied to the land. They can—in theory—freely choose to leave (i.e., choose not to be "repatriated.") In reality, few can withstand the consequences! (Just like wage earners can freely choose a health care plan other than the one(s) offered by their company—they simply cannot afford to do so.)

Finally, to press the unfortunate analogy one step further, what occurs when health care professionals choose to make it a policy to defer to "repatriation"—when initiated and demanded by health care organizations rather than patients—is not, in fact, the "repatriation" but, rather, the "*deportation*" of patients.

THE INTEGRITY OF MEDICINE AND THE NATURE OF THE HEALING RELATIONSHIP

The health care professions, like other professions, are socially created institutions. That is, they are the products of a long-standing social contract: in return for special privileges, health care professionals are charged with special responsibilities and, for the most part, these privileges and responsibilities remain tacit rather than explicit. However, it is quite productive, on occasion, to re-visit those tacit beliefs and expectations that lie behind such institutions and to examine how well they stand up to scrutiny.

For example, contrary to popular opinion, the role of a physician is not—nor has it ever been—simply to cure disease or minimize dysfunction.[7] Rather, the goals of medicine are many and diverse and—depending on the particular blend of physician expertise, patient need and societal expectation, acceptance and commitment to resources—may include (minimally) any and, often, all of the following:

1. To educate self, peers, co-workers, patients, public
2. To relieve pain and suffering
3. To provide comfort to the hopelessly ill and dying

4. To maintain/restore function
5. To minimize dysfunction
6. To cure disease
7. To support and or pursue research

For over two thousand years in the western world, medicine has been a profession that, whatever else it has offered, has dedicated itself to the Hippocratic ideals of beneficence and non-maleficence. The former has generally been reflected in the profession's pursuit of "the patient's good"—however variously it's been construed along the way—and the latter in the profession's cautionary motto: "Above all, do no harm." It is to these two ethical ideals—beneficence and non-maleficence—that the profession invariably appeals in articulating and justifying the means and ends of medicine.

Arguably, the pursuit of these two ideals is a fitting objective for any profession, since every mature profession holds at least three characteristics in common:

1. Specialized expertise unavailable to laypersons
2. External recognition of that specialized expertise (e.g., licensing by the state)
3. Internal control over what is researched, what is taught and what is practiced (viz., the authority to define the limits of its subject matter and expertise, to structure the education and certification of its initiates, and to establish safe standards of practice and adequate disciplinary mechanisms)

In return for allowing professions such exclusive control, the public expects to be reassured that the profession, as a whole, can be trusted not to abuse the special powers and privileges entrusted to them. Along with this expectation, the public assumes—on the basis of this reassurance—that professionals will comport themselves with integrity and fidelity by representing, first and foremost, the best interests of those who seek their assistance. Traditionally, the four pre-eminent professions are teaching, religious ministry, law and medicine. The public recognizes and entrusts these professions, respectively, with the educational development, spiritual counsel, criminal and civil representation, and medical needs of its members. As with each of the other professions, the relationships between medical professionals and patients are, of course, multifaceted. Each profession has its own characteristic set of ethical, legal and social considerations depending on its particular context and subject matter. Likewise, each of these professions includes, under its social considerations, the economic arrangements under which its services are customarily rendered.

As we noted in our introduction, until quite recently the default assumption has been that patient/healer relationships are dyadic, making the responsibilities generated by that relationship dyadic as well. Thus, economic responsibility flowed from patients to health care professionals. How that responsibility was discharged in any given case—whether directly by the patient or indirectly through some third party representing the patient—was still, in the final analysis, the personal responsibility of the patient. Likewise, responsibility for health care expertise flowed from the profession to its patients. And again, how that responsibility was discharged in any given case—whether directly by the physician or indirectly through the various members of the health care team—was ultimately the personal responsibility of the physician as a professional.

This default assumption about the personal responsibilities between patients and physicians has remained fairly stable across societies, irrespective of their particular economic and political structures—even with the introduction of "managed care" here in the United States. There are several reasons why this is the case. First, professional relationships in health care are unique in at least two respects:

1. They often occur in a context of dramatic intensity often not experienced in other social associations
2. While they are exchange relationships, they require intimacy between persons who, under any other circumstances, are not intimate, and may even be complete strangers

Such conditions render both patients and physicians vulnerable in rather specific ways. Patients risk vulnerability whenever they submit their private lives to the scrutiny and manipulations of physicians, and physicians risk vulnerability insofar as their best efforts may be rejected or challenged by patients, peers or society. A deep and abiding mutual respect and responsibility—reinforced by the ideals of beneficence, non-maleficence and justice—has been the tried and true counterbalance to the risks of vulnerability faced by both parties to the patient/physician relationship.

The second reason for this "default" assumption is that social attitudes and practices are, like all habits, slow to change and hard to break, even once it is evident that they have become useless or even counterproductive. This second reason helps to explain why patients and physicians continue to behave as though their commitments and obligations have not changed when, in fact, the relationship has undergone a significant and fundamental restructuring under "managed care." The patient's standing within the relationship is compromised by the fact that patients today have little, if any, control over, representation or over-sight with regard to how their health care is chosen or funded. However, it is the physician's standing—both as individual practitioner and as professional—which is most critically challenged.

The individual physician's loyalties have become divided between the traditional responsibilities of beneficence and non-maleficence towards patients and the novelty of contracting with or being employed by (and thus having an exclusive and separate fiscal responsibility to) a new, third party to the relationship whose interests all, eventually, reduce to an entrepreneurial bottom line.

The third identifying feature of a mature profession mentioned earlier—the internal control that allows a profession to define its subject matter and expertise, to structure the education and certification of it's initiates and to establish safe standards of practice and mechanisms by which to discipline it's members—is under direct threat by managed care. Giving PPO's the power to re-shuffle patient/physician relationships with the impunity they currently possess allows them to maximize what will be most fiscally beneficial to the managed care organization while seriously undermining the very basis upon which the profession of medicine, as a social institution, rests—the triad of *practice*, *education* and *research*. And these three elements are inseparably linked: whatever affects one eventually affects them all; whatever interrupts one eventually interrupts all.

Such power over the healing relationship is, by far, the most serious threat to the integrity of medicine since, as of this moment, there is nothing to motivate the participants of today's managed care organizations to consider anything or anyone other than their own best interests—and these as narrowly conceived as possible. When there is no motivation to continue to cultivate a habit or virtue, it will eventually be replaced by the cultivation of more fruitful habits and virtues even while it continues to be praised as a worthy, albeit empty, ideal.[8] Research and on-going education are, as it were, social habits. That is, they have been—either tacitly or explicitly—recognized by the public as necessary elements of maintaining the integrity of the profession. Unless we value research and on-going education enough to incorporate them in daily practice and reward those who practice such habits—and not simply pay lip service to them as distant ideals—they will die and the profession will stagnate. Experience has taught us what happens when research and education are considered less important than practice: at best, the practice is transformed into an empty, harmless (and ultimately rather useless) symbolic social ritual; at worst, it can be diverted from a public to a private good, serving only the narrow—and very short-term—interests of a powerful few.

THE MANAGED CARE ENVIRONMENT AND THE THREAT TO PATIENT CARE

The rationale employed to justify the rise of HMO's and PPO's is, basically, an economic one: health care has become too costly. Of course, there are other, non-economic "costs." But, when the emphasis—for whatever reasons—remains focused on economic costs to the near exclusion of all else, health care is discussed from within—and, in crucial ways, defined by—the particular economic framework adopted by the society in question. In the United States, unlike most other democracies in the world, the economic framework for the exchange of virtually all goods and services is the open, competitive marketplace.

As HMO's respond to the mechanisms of the marketplace, they are under increasingly stronger pressure to further reduce their costs. This, in turn, compels HMO's to transfer their patients either into hospitals owned by the HMO itself or into facilities that offer the best contracted rates. Facilities that concentrate their efforts solely on practice do not incur—or have to absorb—the additional expenses associated with both research and teaching programs. For example, since many trauma centers are located within a university setting, they will have a broader—and, hence, relatively more expensive—focus precisely because they are committed not simply to practice outcomes, but to research and teaching outcomes as well. When the driving forces in health care are overwhelmingly entrepreneurial what mechanisms, in such a system, will value and preserve the necessary conditions for good practice—namely, ongoing research and education?

Patients caught in the midst of this competitive battlefield are forced to give up whatever trusting relationship has developed with the health care team initiating their care and to re-establish this important connection with a completely new set of health care team members. This can produce significant disruptions, psychological as well as physical. (One of the classic ways to make such disruptions more palatable by those perpetrating them is, of course, through the deliberate misuse of language, a concern we will address more fully in the next section.)

Clearly, it is beyond the scope or aims of this paper to reiterate the usual philosophical arguments in defense of the marketplace and, fortunately for our purposes, it is not necessary. To press our case we need only to challenge several key presuppositions about the marketplace that are widely accepted as uncontroversial in order to reject the competitive marketplace as a suitable economic framework for the exchange of certain basic kinds of goods and services such as education and basic health care services.[9]

CHALLENGING THE PRESUPPOSITIONS OF THE MARKETPLACE

The success of the marketplace depends upon a broad base of consumers who (1) have sufficient funds to enter and compete in the market, (2) are sure of what they want and need, (3) are able to judge quality and price according to a standard, and (4) have sufficient time to deliberate, compare and "shop around."[10] The authors readily concede that, generally speaking, when such minimal conditions are met, not only will the marketplace function well but it will tend to benefit all. However, it is far from clear that any of these minimal conditions can be met with regard to the commodification of health care. Empirical evidence clearly refutes the first presupposition: That is, (1) a large and growing segment of our population here in the U.S. does not, in fact, have sufficient funds to enter and compete in the health care market. Moreover, the second and third presuppositions beg the question since the very points at issue are whether consumers can, in fact, (2) make appropriate determinations about their health care needs and wants and (3) judge the quality and price of health care. Lastly, it is entirely unrealistic to suppose that consumers would (4) have sufficient time to "shop around" for the "best" health care "deal" they could get—even if they could make these appropriate determinations.[11] The empirical data strongly suggest that the implicit "*caveat emptor*" strategy of the marketplace—viz., "let the buyer beware"—is clearly an inadequate method of exchange for at least some social goods and services.

CONCLUDING REMARKS

The practice of "repatriation" exemplifies two of the most unsettling "managed care" issues for health care ethics today: the ethical challenges that physicians face in meeting their obligations towards each patient who enters a healing relationship with them and the crisis that the profession of medicine itself faces as a social institution. The response thus far—from patients and physicians alike—has been consistent with our naive predilection towards a philosophy of rugged individualism: "What can I do?" By framing the question in this way we not only guarantee frustration, but court defeat, since an individualistic approach to what, in reality, are "system errors" can offer only piecemeal, symptomatic treatment of a deeper, systemic problem that such an approach can only begin to appreciate.

We are not denying that change must begin with individuals; of course it must. However, the issues we have been discussing cannot be resolved by individuals working in isolation from and, too often, at cross-purposes with, one another. Systemic problems require systemic—i.e., political—resolution. It is not unreasonable to expect that, in a democracy, systemic problems will be publicly addressed and the broad outline of the mechanisms for their resolutions publicly crafted. And, while democracy is synonymous with the idea of representation, it is representation of a kind that embodies a clear understanding of the necessary interdependence of individual and community—unsullied by the demeaning posturing and pandering to special interests that we must endure from our political "representatives today. After all, the development of individuality presupposes a flourishing community dedicated to nourishing the unique interests and strengths of all of its members; just as the development of a flourishing community cannot even be envisioned without the strength of talented and skilled individuals.[12] There are multiple and overlapping ways such representation can occur, depending on the number and nature of social roles an individual has accepted or is expected to play.

As a society, we give to the four major professions mentioned earlier wide latitude in representing our best interests. Physicians wear, for example, at least three "hats" because of who they are, the roles they play, and their position in society. First, as practicing experts, physicians are obligated to patients with whom they form professional relationships. Part of what that obligation entails, practically speaking, is being willing—and able—to represent with fidelity the individual patient's best interests as they are determined jointly within the patient/physician relationship.

Second, as members of a socially supported and sanctioned profession, physicians are obligated both to the profession itself and to the society that recognizes and supports the profession. Again, part of what that obligation entails is representation: helping both individual patients and the public at large to understand the potential of the patient/physician relationship, what it can and cannot offer and providing a unique perspective on issues—their antecedent conditions as well as their material consequences—that affect or are relevant to the profession and to health care (individual and communal). However, another equally important part of that obligation is to preserve the integrity of the profession by reminding the public how necessary continuing research and education are to the health and integrity of the profession itself.

Third, as citizens of a democracy, physicians are obligated as members of a community. A large part of what that entails includes contributing the unique perspective that their particular experience has afforded them to the public dialogue that is so essential for a healthy democracy. Who else is better qualified to articulate most clearly some of the central questions that the public needs to be asking: e.g., whose best interests are served under the current managed care organizations? Who do HMOs and PPOs represent? To whom are they responsible?

In a democracy, social institutions are structured—and expected—to benefit all. When these institutions become out-dated and less effective, as they invariably must over time, they need to be re-constructed. When the public fails to hold itself responsible for making those readjustments (either directly or through its social and/or political representatives), it creates a vacuum that narrower, private interests are sure to fill—characteristically in unrepresentative ways.

Medicine is a social institution. If we, as a democratic public, are committed to a viable reconstruction of a large aspect of medicine—namely, equitable access to and delivery of health care—we must recognize and take seriously its place in the larger institution as a whole. This includes understanding the difference between what is essential for medicine as a social institution to survive and flourish and what is not, how it interconnects with other social institutions and, most importantly, the values and character traits it simultaneously reflects and fosters in both its practices and the ideals it represents for society. Acquiring such understanding in a modern democracy requires the public to encourage and nourish the development of experts and professionals (and in all fields, not simply medicine) who have integrity, who are committed to contributing their respective knowledge and perspectives to the on-going, public dialogue and debate that characterize healthy, robust and democratic institutions.

ENDNOTES

1. To those critics who attempt to explain or excuse the democratic inequities of the United States by arguing that, technically speaking, we a republic, our only comment is that, technically speaking, so was Nazi Germany—thus, an appeal to the term, "republic," does little to explain, and even less to excuse, a state's behavior. The term, "republic," is extremely vague and uninformative and, according to most experts in the field, remains quite controversial for this very reason. In its more archaic form, it simply referred to the state or "common weal." Its current meaning is equally uninformative: it is considered by most government scholars to be simply "a state in which the supreme power rests in the people and their elected representatives or officers, as opposed to one governed by a king or similar ruler." (*Compact Edition of the Oxford English Dictionary,* London: Oxford University Press, 1984)
2. Source: Health, United States: 1999, "www.cdc.gov/nchs/fastats/hinsure.htm", (website of the United States Department of Health and Human Resources, Center for Disease Control).
3. Like all forms of insurance, health care insurance is based on the notion of shared risk, whereby those facing a common vulnerability (in this instance, disease and disability), can choose to pool their resources for a relatively small fee. Thus, if a participant then becomes diseased and/or disabled, financial assistance is made available.
4. We have purposely used the broader term, "respect," instead of the separate terms "autonomy" and "justice" since "respect between persons" already entails a mutual rec-

ognition of the other's autonomy and the necessity of working out a fair and equitable way of dealing with each other. In this way, we emphasize the living and dynamic interrelationship between autonomy and justice, rather than treat them as pre-existing and competing claims that are brought into a relationship. For a sustained argument for a homeostatic reconstruction of current dogma surrounding the notions of autonomy, beneficence, non- maleficence and justice (i.e., the "Georgetown mantra"), see Roberta Springer Loewy, *Integrity and Personhood: Looking at Patients From a Bio/Psycho/Social Perspective*, (Ann Arbor, MI: UMI Dissertation Service), 1997.

5. For a brief, informative summary of the history and ideology behind health care insurance and the financing of health care, see Christine K. Cassel, "Health Care Financing: Introduction," *The Encyclopedia of Bioethics*, ed. by Warren T. Reich (NY: Simon & Schuster Macmillan, 1995), 2:1049- 57, esp. 1051.
6. It is far beyond the scope of this paper to defend those criteria traditionally considered definitive of a profession. For a more thorough discussion of this topic, one must begin with some of the classic literature of the sociology of profession. In medicine, see, for example, Eliot Friedson *Profession of Medicine*, (NY: Dodd, Mead), 1970; *Moral Responsibility and the Professions*, ed by Bernard Baumrin and Benjamin Freedman, (NY: Haven), 1982; Paul F. Camenish, *Grounding Professional Ethics in a Pluralistic Society*, (NY: Haven), 1983.
7. This viewpoint is not simply wrong (e.g., most plastic surgery today is done for aesthetic reasons and not for the purpose of restoring function), it also reinforces an unfortunate stereotype, one that portrays patients as submissive, passive recipients of a special technical expertise that only physicians can perform. Such a view threatens to return medicine to the status of primitive, mystical art, the physician to shaman and the patient to faithful—and fearful—supplicant. Such a view threatens all of the sciences today—a crisis of grave proportion that can be overcome only by critical public dialogue about the means and ends of the various sciences as they exist within the context of the particular society underwriting them. For an eloquent defense of the methodology of science and scientific thinking against the rise of a pseudoscience that is nearly religious in its fervor, see Carl Sagan's *The Demon-Haunted World: Science as a Candle in the Dark* (New York: Ballantine Books), 1996.
8. One of John Dewey's constant laments was that we fail to subject ideals to the same open, tolerant, but critical intellectual process used to evaluate the problems, issues and practices of daily living: "Men [sic] hoist the banner of the ideal, and then march in the direction that concrete conditions suggest and reward." *The Quest for Certainty: The Later Works*, 1925-53, ed. by Jo Ann Boydston, (Carbondale: Southern Illinois University Press, 1988), Vol. 4:1929, pp. 224-5. While Dewey fully agreed that one of the criteria for critical thinking is the ability to make distinctions (such as the distinction between real and ideal or that between theory and practice), he insisted that none of our ideas—including the fine distinctions we make—should ever be immune from critical scrutiny, scientific examination, and subsequent reconstruction. Unless ideals are tested in the crucible of experience to see whether and how they can improve the actual conditions of human existence, they will become formal, but empty, symbols.
9. To describe and critique the assumptions of the marketplace we draw from the work of Erich H. Loewy. See, for example, his article, "Of Markets, Technology, Patients and Profits," *Health Care Analysis*, vol. 2: 101-09 (1994).
10. Ibid., p. 105.
11. Ibid., pp. 102-04.
12. Ibid., p. 108.

Chapter 12

Rationing Health Care in the United States and Canada

Walter Glannon
Assistant Professor
Biomedical Ethics Unit, McGill University
Montreal QC H3A 1W9, Quebec, Canada
e-mail: glannon@falaw.lan.mcgill.ca

Key words: health care systems, health care allocation, fairness, efficiency, rationing, queuing, managed care, Oregon Basic Health Services Act

Abstract: Any health care system should be grounded in the ethical principle of fairness in access to services and the economic principle of efficiency in allocating resources. In this paper, I explore the relationship between ethics and economics in possible two-tiered rationing schemes in American and Canadian systems. In the US, rationing through managed care occurs in the form of constraints on the type and number of services doctors perform. In Canada, rationing occurs in the form of queuing, with comparatively long periods of waiting time for surgery and other treatments. Managed care organizations in the US have temporarily increased efficiency in the delivery of care by reducing waste and unnecessary services, but not fairness, as evidenced by the 44 million uninsured and underinsured. I consider whether adopting a universal model along the lines of the Oregon Basic Health Services Act might ensure both fairness and efficiency. On this model, guaranteeing that all Americans had a decent basic minimum of health care would mean excluding some expensive treatments from health plans, though people with the ability to pay for these treatments might have them. Such a two-tiered system would be fair provided that the lower tier entailed a decent minimum that met people's basic health care needs. Tiering in the Canadian system would involve allowing those with the ability to pay for expedited care to jump the queue and thereby cut their waiting time. This might ameliorate the problem of waiting for people who cannot afford to jump the queue because they are unable to pay for expedited care.

Changing Health Care Systems from Ethical, Economic, and Cross Cultural Perspectives, edited by Loewy and Loewy. Kluwer Academic/Plenum Publishers, New York, 2001.

The present Canadian system is fair but not always efficient in the delivery of services. In the US and Canada, a two-tiered system might be the most viable way to allocate health care resources both fairly and efficiently.

Any health care system should be grounded in the ethical principle of fairness and the economic principle of efficiency. It should ensure that all people with medical needs have access to a decent minimum of care and that medical treatments be beneficial to patients while falling within certain cost constraints. To be viable, health care systems must conceive of and realize universal access to medical services and cost-effectiveness as complementary goals.

Unfortunately, fairness and efficiency often pull in different directions. Thus, ensuring the complementarity of these principles in a health care system is no easy task. The problem is especially acute in the United States, where roughly 44 million Americans are uninsured and therefore a significant portion of the population has no access to any health care. While the American health care system has become more efficient in the last decade due largely to the ability of managed care to reign in costs, it remains fundamentally unfair to too large of an underserved population. In Canada, by contrast, since the formulation and implementation of the Medical Care Act of 1966, the health care system has been comparatively fair and efficient. However, an aging and sicker population, plus reduction in the transfer of federal funds to the provinces following the Expenditures Restriction Act of 1991, have led to increasing claims on limited elective and urgent services. Consequently, both the fairness and efficiency of the Canadian system are being threatened.

The most promising—perhaps the only—way to reconcile fairness with efficiency is through a particular form of medical rationing. In fact, health care in the United States already is rationed by price and ability to pay, which has the ethically objectionable consequence that many people have no access to health care because of their inability to pay for it.[1] Health care also is rationed in Canada in the form of queuing, where people often wait considerably long periods for elective and, increasingly, urgent medical interventions. Longer waiting time for both elective and urgent procedures has the ethically objectionable consequence of an unacceptably high number of people having to suffer longer from diseases or disabilities and worse health outcomes. The problems unique to the distinctive forms of rationing in the United States and Canada illustrate the need for an alternative form of rationing in each case. With this need in mind, I will propose and explore the idea of a two-tiered health care system, where a publicly funded tier guarantees access to a decent basic minimum of care and where a privately funded tier consists in supplemental, or expedited, care for which individuals pay. I will discuss how this would function in American and Canadian health care systems, respectively. Then I will address the main objection to tiering, namely, that it results in inequalities that are unfair and hence morally objectionable because it makes access to health care based on ability to pay rather than medical need.

Fairness is concerned with meeting the claims of need of different people. It requires that claims be met in proportion to their strength, where strength is a function of degree of need.[2] Accordingly, given a general scarcity of resources, a fair allocation will be one in which people with the strongest claims of medical need are given priority in having their needs met. Fairness requires that the more urgent needs of the worst off be given priority over the less urgent needs (or preferences) of the better off in terms of health status. For the issue at hand, the worst off members of society are those with the poorest quality of life combined with the shortest life expectancy. The majority of these are the socio-economically poor, who not surprisingly constitute the majority of the uninsured.

Fairness is one aspect of justice, and a just health care system is one that gives priority to the needs of the worst off. The underlying ethical and political rationale for this view is Rawls' Second Principle of Justice, also called the "Difference Principle" or "Maximin Rule."[3] This says that the only admissible inequalities are those that work to the benefit of the least advantaged members of society. Significantly, while this is indeed an egalitarian (as distinct from a libertarian or utilitarian) conception of justice, it does not imply that in health care equality as such has intrinsic value. Rather, it says that in allocating scarce medical resources, priority should be given to the worst off so that they will have an equal opportunity for access to a decent minimum of health care. In turn, they will have an equal opportunity with others for achieving a reasonable level of well-being over the course of their lives. More precisely, what matters is not so much how the health status of the worse off (sick) compares with that of the better off (healthy), but instead how the worst off fare with respect to an absolute baseline of care where their basic medical needs can be met.[4]

Still, there are two senses of "worse off" which need to be distinguished. One pertains to health status at particular times, while the other pertains to health status over long periods of time. Some people are worse off in terms of acute urgent conditions due to trauma or infection, while others are worse off in terms of chronic conditions like asthma, diabetes, and heart disease. But if we try to accommodate both types and attempt to meet everyone's medical needs without evaluating degrees of urgency as well as outcomes of treatment, then costs are likely to become intolerably high. Hard allocation choices must be made.

No liberal democratic society can afford to have health care constitute a disproportionate amount of GDP compared with other social goods, such as education, housing, or environmental protection. For most people, health care is more important than these other goods. But this cannot be assumed, as many would hold that an educated public is as socially important as a healthy one. With respect to health care, there are practical political limits to how willing the general public will be to pay higher taxes, or how much of a financial burden employers would be willing to take on in providing health care for their employees. Furthermore, increased government spending on health care, without corresponding cuts to other social goods, would add to the national debt. This would be ethically objectionable because of the unfair burden it would impose on future generations. Hence the need for controlling health care costs through rationing. But what form should rationing take?

It is worth emphasizing that rationing is necessary to both ensure universal access to decent basic care and to control costs. A just health care system, one with universal access to a basic minimum, would include such things as immunizations, prenatal care, antibiotics, insulin for diabetes, emergent care, and continuity of care with an overseeing primary care physician. In the United States, it would require significant government expenditures providing this basic package to the poor. Beyond this, a combination of employer and personal contributions would provide the remainder of the needed money for care for everyone. Providing all people with a decent basic minimum of care would satisfy the justice requirement.[5] This would constitute the primary tier of care.

But because costs must be taken into account, treatments that failed to maintain patients at or restore them to a baseline of adequate physical and mental functioning for an extended length of time, or which raised their functioning well above the baseline, should not be funded either by the government or by private insurers. These would fall into a secondary tier of care. Patients would be allowed to have these treatments on the condition that they pay the real cost. These would include enhancement treatments, such as cosmetic surgery, and assisted reproduction technologies, which do not meet people's basic medical needs. More controversially, individuals (or their families) would have to pay for costly interventions that offer little therapeutic benefit. These would include dialysis beyond a certain age, many life-prolonging treatments in the ICU, and bone marrow transplants for advanced metastatic breast cancer.

The point here is that rationing through tiering would be a strategy for controlling costs provided that universal access to basic care already had been guaranteed. Cost-effectiveness in the evaluation of health outcomes for certain procedures would be a necessary complement to universal access within this framework. Despite its imperfections, the main principles and prioritization in the Oregon Basic Health Services Act of 1989 capture the essential features of my proposal. This Act is consistent with the Rawlsian theory of justice in general and the worst-off priority principle in particular. Something along the lines of this plan, writ large, would be the most promising way to reconcile fairness with efficiency in American health care.

Health care spending in the United States is the sum total of several disparate spending sectors—Medicare, Medicaid, managed care, and private non-managed care. The Canadian health care system is a social democratic health insurance system with publicly funded universal access to coverage for a fairly generous set of benefits.[6] This was mandated by the federal Medical Care Act of 1966. By 1971, all provinces had universal medical and hospital services in insurance plans eligible for federal cost sharing. Today, health care in Canada is controlled mainly by the provinces. Currently, Canada ranks fifth among OECD countries in percentage of GDP from health care (9.3%) and lags rather far behind the United States (14.0%) in this regard. This is because certain structures within the Canadian system are better able to control costs. First, because there are no private insurers, no costs of estimating risk status in order to set differential premiums, and no allocation for shareholder profits, Canada's single-payer system has less administrative overhead than the American system. Second, there is a different specialty mix, with fewer specialists and more primary care physicians, the converse of what exists in the United States. Third, physician fees are lower in Canada. Fourth, Canada utilizes much less intensive technology for diagnosis and treatment, such as MRI and CT scans, largely by limiting the supply of such technology.

The combination of lesser availability of intensive technology, fewer specialists (radiation oncologists and anesthetists, in particular), fewer nurses, an aging population, and reduced transfers of funding from the federal government to the provinces has meant that Canadians have been waiting longer for medical services. Such queuing has been the Canadian method of rationing scarce health resources. Until recently, queuing was largely confined to non-urgent, non-emergent elective procedures such as hip replacement surgery. Now, however, many patients have to wait an unacceptably long time for not just elective but also urgent and emergent care. The problem is especially acute in Quebec and Ontario, where many patients have to wait up to 28 weeks for radiation cancer treatment. This has adversely affected not only patients' quality of life but quantitative health outcomes as well. Furthermore, in all provinces emergency medicine has been seriously affected by a shortage of nurses. In attempting to resolve some of these problems, Quebec and Ontario provincial governments have permitted patients to receive more timely access to more widely available MRI diagnostic technology and radiation cancer treatment in such American cities as Buffalo, Cleveland, and Burlington, VT, with the Canadian provincial governments footing the bill. This has resulted in higher costs to Canadian Medicare than what would have been the case if these patients had received timely treatment in Canada.

Tiering involving *limited* privatization would be a more cost-effective way of resolving the problem of dangerously long waiting periods for necessary treatment. Indeed, this already has been suggested by different voices in Quebec and Alberta, and Alberta has initiated a plan (Bill 11) that would include a network of private hospitals where patents would pay for private services for elective surgical procedures. With appropriate regulation by the federal and provincial governments, there could be a limited Canadian-based private tier with more readily available technology and more specialists performing expedited procedures that people would be able to pay for with their own money. The long-term strategy would be to train more specialists within Canada, some of whom would practice in the public tier, others in the private tier, depending on certain agreements made at the start of their medical education. One of these agreements might be higher tuition and fees for those intending to practice in the private tier. This would be in violation of the Canada Health Act of 1984, which forbade "extra-billing" and "user fees" of any sort, in which case significant and politically unpopular changes would have to be made within the existing health care system. But without some accommodation of private care within the existing public system, Canadian Medicare may very well be threatened by financial crisis at some point in the near future.

This suggests that access to some forms of health care would be based on ability to pay rather than on need. And because some people are financially better off than others, such a system would seem to be unfair to those who could not afford to pay for expedited care. Yet, examined from a more long-term macroallocation perspective, if some people were able to jump the queue and thus cut their waiting time in half, then this would not only reduce the risk of morbidity or even mortality by not having to wait so long for treatment. It also would ameliorate the problem of waiting for people who cannot afford to jump the queue because they cannot afford to pay for expedited care. The quality of care for all would improve.

On the other hand, one could argue that creating a private tier would lead to the gradual erosion of the quality of care in the public tier and threaten the very idea of a decent basic minimum of health care. Attracted by economic incentives, better actual and would-be physicians might gravitate to the private tier, especially if the general perception among physicians were that the government was underfunding health care and not paying them their fair share for their services. It is important to point out, though, that justice and fairness pertain to access to a decent basic minimum of health care, not to supplementary or expedited care above this critical level. If the emergence of a private tier undermined access to a decent minimum in the public tier, then creating a private tier would be ethically objectionable. But it would be objectionable because of the unfairness to those in the public tier, not because of inequalities in more or less timely access to care. What would make tiering unfair is not that it would make those with access to only the queued public tier worse off in *relative* terms compared with the better off who have access to the expedited private tier, but that the decent minimum itself would erode. Consequently, those in the public tier would be worse off in *absolute* terms regarding the basic minimum. Any inequalities between the better and worse off would not necessarily imply unfairness to the latter regarding access to basic care. And if these inequalities were not unfair, then they would not be ethically objectionable.

Daniel Callahan has argued that "rationing is likely the only way in which we can improve care for the poor *and* manage our health care system in a more efficient manner."[7] This is in reference to the American health care system, and I have proposed a model of rationing involving a two-tiered structure to guarantee universal access to a decent basic minimum of care while controlling costs. I also have proposed a two-tiered model of rationing as a solution to the Canadian problem of patients waiting so long for urgent medical procedures that health outcomes are seriously compromised. While being attentive to the differences between the two countries' health care systems, in each case the aim of the model is to reconcile fairness with efficiency. As a matter of justice, universal access to a decent basic minimum of care is required. Again referring to the American system, in Larry Churchill's words: "A health care system which neglects the poor and disenfranchised impoverishes the social order of which we are constituted. In a real and not just hortatory sense, a health care system is no better than the least well-served of its members."[8] As a matter of economics, cost-effectiveness must figure in considering treatments. Since no health care system can be all things to all people, since it cannot meet all of their needs and preferences, rationing through tiering is necessary if we are to meet people's needs the best we possibly can.

To be sure, the type of model I have proposed would be met with resistance. In the United States, many libertarians would argue that it would be unfair for some to subsidize the health care of others through higher taxes in order to ensure universal access to basic care. In Canada, many egalitarians would argue that any privatization of health care would be anathema to the principles of equal access to and public funding of health care, which largely form the core of Canadian identity. But the problems of 44 million uninsured Americans and of Canadians having to wait unacceptably long periods for urgent treatment require fundamental changes in their respective systems. Adherence to political ideology is not likely to be very helpful. Rationing through tiering may be the most viable way to achieve the desired ends of universal timely access to and cost-effectiveness in the delivery of health care.

ENDNOTES AND REFERENCES

1. Daniel Callahan, "Rationing Health Care: Social, Political, and Legal Perspectives," *American Journal of Law and Medicine* 18 (1992): 1-16. Uwe Reinhardt, "Reforming the Health Care System: The Universal Dilemma," *ibid.*, 19 (1993): 1-20.
2. Michael Lockwood, "Quality of Life and Resource Allocation," in J.M. Bell and S. Mendus, eds., *Philosophy and Medical Welfare* (Cambridge: Cambridge University Press, 1988): 33-56. John Broome, "Goods, Fairness, and QALYs," *ibid.*: 57-74; John Harris, "More and Better Justice," *ibid.*: 75-96.
3. John Rawls, *A Theory of Justice* (Cambridge, MA: Belknap Harvard University Press, 1971): 7-11, 40 ff., and "Social Unity and Primary Goods," in B. Williams and A. Sen, eds., *Utilitarianism and Beyond* (Cambridge: Cambridge University Press, 1982); 159-185.
4. Derek Parfit, "Equality or Priority?", Lindley Lecture, University of Kansas, 1995. Thomas Nagel, *Equality and Partiality* (Oxford: Oxford University Press, 1991).
5. Allen Buchanan, "The Right to a Decent Minimum of Health Care," *Philosophy & Public Affairs* 13 (1984): 55-78.
6. John K. Iglehart, "Canada's Health Care System Faces its Problems," *New England Journal of Medicine* 322 (February 22, 1990): 562-568, and "Revisiting the Canadian Health Care System," *New England Journal of Medicine* 342 (June 20, 2000): 2007-2012; C. J. Tuohy, *Accidental Logics: The Dynamics of Change in the Health Care Arenas in the United States, Britain, and Canada* (New York: Oxford University Press, 1999); M. Kennedy, "73 Percent Back Private Health Care: Most Canadians in Favour of Two-Tiered System if it Means 'Timely Access' to Care, Survey Finds," *Ottawa Citizen*, January 22, 2000.
7. Daniel Callahan, "Rationing Health Care: Social, Political, and Legal Perspectives," *American Journal of Law and Medicine* 18 (1992): 4.
8. Larry Churchill, *Rationing Health Care in America* (Notre Dame, IN: University of Notre Dame Press, 1987): 95 ff.

Chapter 13

Altering Capitation to Reduce the Incentive to Undertreat Patients Inappropriately

Rory Jaffe
Chief Compliance Officer, UC Davis Health System
Associate Medical Director, UC Davis Medical Group
Associate Professor, Anesthesiology and Pain Medicine
University of California, Davis
Sacramento, California 95817
e-mail: rsjaffe@ucdmc.ucdavis.edu

Key words: health care systems, socio-economic determinants of health care, capitation, clinical practice, internalized norms

Abstract: Capitation, the payment of a fixed monthly fee for covered people, is being used by insurers to eliminate the financial incentive for over-provision of services that is present in fee-for-service plans. Physicians typically respond by forming groups to manage the financial risk associated with capitation. These groups institute methods of sharing the "cap" dollar among the involved primary care practitioners and specialists. To reduce utilization groups commonly "subcap" each specialty, giving them fixed payments per month regardless of the amount of care they provide to the covered members. Thus, individual physicians are rewarded with a payment per unit of work that rises towards infinity as work declines towards zero. This inappropriate incentive is counterbalanced by the internalized norms we assume all physicians gain as part of their professional education and experience. Unfortunately, it appears that this financial incentive can overwhelm these internalized norms in a number of physicians to the point of not only reducing care below levels associated with what an individual patient would regard as appropriate, but also below levels associated with maximal societal gain. In this paper I present a compensation method that can be adjusted to provide a financial incentive to reduce services when they seem to be excessive, while reducing or eliminating the incentive to reduce services when utilization is already low. It uses a payoff

Changing Health Care Systems from Ethical, Economic, and Cross Cultural Perspectives, edited by Loewy and Loewy. Kluwer Academic/Plenum Publishers, New York, 2001.

function to adjust fee-for-service payments based on utilization. The function allows maximal incentive (measured as the derivative of the function fee [utilization]) to be placed where it is judged most appropriate, with little additional incentive at already-low utilization levels.

1. INTRODUCTION

Capitation, the payment of a fixed monthly fee for covered people, is being used by insurers to eliminate the financial incentive for over-provision of services that is present in fee-for-service plans. Physicians typically respond by forming groups to manage the financial risk associated with capitation. These groups then institute methods of sharing the "cap" dollar among the involved primary care practitioners and specialists.

To reduce utilization, groups commonly will "sub-cap" each specialty, giving them fixed payments per member per month regardless of the amount of care they provide to the covered members. Given a fixed dollar amount, individual physicians are rewarded with a payment per unit of work that rises towards infinity as work declines towards zero. This inappropriate incentive is counterbalanced by the internalized norms we assume all physicians gain as part of their professional education and experience. Unfortunately, it appears that this financial incentive can overwhelm internalized norms in a number of physicians, to the point of not only reducing care below levels associated with what an individual patient would regard as appropriate, but also below levels associated with maximal societal gain.

The quest, then, is to develop a compensation method that can be adjusted to provide financial incentive to reduce services when services seem to be excessive, while reducing or eliminating the incentive to reduce services when utilization is already low. This paper presents one such method, using a payoff function to adjust fee-for-service payments based upon utilization. The function allows maximal incentive—measured as the derivative of the function fee (utilization)—to be placed where it is judged most appropriate, with little additional incentive at already-low utilization levels.

2. HEALTH CARE ECONOMICS

Understanding capitation and its purpose requires some knowledge of the reasons this payment method recently became popular in the USA.

In the USA, health care costs began rapidly rising in the 1960's, and have continued to increase consuming, today, nearly 1/7th of our gross domestic product. This rise coincided with the initiation of government programs to extend health insurance to those formerly unable to receive other than charity care.[1] While this was an unforeseen consequence of Medicare and Medicaid, looking at market economics readily explains it.

In an ideal market, individuals only purchase those products and services priced at or below the value that they, as consumers, place upon them. Producers respond to this demand and price sensitivity by making as much as possible as long as cost is not more than the price they can get. At equilibrium, cost is at the minimum possible, and price is equal to the economic cost. So price, value, and cost "meet." This "invisible hand" results in the goods and quantity society wants at the price society is willing and able to pay.

> As every individual endeavors to employ his [effort] so that its produce may be of the greatest value, every individual necessarily renders the annual [surplus] of the society as great as he can. He neither intends to promote the public interest, nor knows how much he is promoting it, [but] he is in this led by an invisible hand to promote an end which was no part of his intention. —Adam Smith[2]

Unfortunately, health care has not functioned well in the open market. Concern about distributional justice is just one of the factors, and was the major impetus behind the establishment of government insurance programs. People not covered by Medicare and Medicaid also desired insurance, as they were unwilling to face the risk of being unable to pay medical costs in a crisis. Unlike other goods, where a missed opportunity to buy can be made up later, being unable to buy health care when needed could lead to irreversible consequences, giving people little choice in determining the timing of their purchases.

Insurance helped address the problems of distribution and emergency needs, but also prevented proper price signaling in the health care market. When an insured person uses healthcare services, he or she pays either a small percentage (traditionally 20%) or, in many managed care plans, even less (such as $5), or nothing. No longer does the price and cost of service have to meet. The patient sees a much smaller price, and values the service accordingly. At its simplest, we can make the assumption that with a 20% co-payment, consumers will be willing to purchase services that cost 5 times as much as the value of the service. Producers, such as physicians, can also price their services well above the both their value and cost, as consumers will be relatively insensitive to price. This market disconnect became one of the main drivers of health care inflation:

Nowadays people know the price of everything and the value of nothing.
–Oscar Wilde[3]

Patients are not necessarily upset by this problem. With traditional insurance, physicians' financial incentives usually are aligned with individual patients' desires. The more a physician does, the more he or she makes. This leads to satisfying patients' basic medical needs as well as their wants. Physicians also had an incentive to produce beyond that needed for the patient's benefit. While as professionals, they have internalized norms with a countervailing effect, there are certainly physicians who do more than is good for their patients. Given the sophistication needed to properly assess the appropriateness of care, patients in traditional insurance plans typically see no conflicts between physicians' motivations and their own.

The payment systems in the USA hid the true cost of health care from consumers, but price inflation did present employers and government with ever-increasing bills. These parties are the main direct payers of health care, and eventually began to resist price increases. As this resistance to premium increases buffeted insurers, insurers began looking for ways to manage the expenses of health care.

3. INSURERS' RESPONSE

Insurers instituted managed care so that they, not physicians, would determine which services were appropriate for a patient to receive. Managed care before capitation looked to health care providers like traditional insurance with one major change: they no longer could do what they wished. Rather, approval was needed for many services. The financial incentive remained to do more (if approved), so the providers and consumers continued to have an incentive to engage in low-value high-cost care. The insurer became the enemy, and had conflicts with providers and patients.

Many insurers have now turned to capitation. Capitation changes the incentives for care and alters the traditional physician-patient alignment of interests. In capitation, a group of providers is given a fixed amount per month to take care of a group of patients (Appendix A, figure 1). This is termed "dollars per member per month", abbreviated pmpm. The group pays for any care given outside the group, and distributes the rest to individual physicians in the group, often on a pmpm basis. The less spent on patient care, the more profit the group makes. The incentive is to minimize expenses. The physician group is now financially aligned with the insurer, not the patient. And with the distribution to individual physicians on a pmpm basis, the individual physicians are motivated similarly.

The effect of capitation is more dramatic when we look at the payment per unit of work. Under fee for service, each additional work unit yields more money. With capitation, however, the money is constant for any level of work. So the compensation per work unit rises dramatically the greater work is decreased (Appendix A, figure 2).

Capitation is attractive to insurers for several reasons. First, insurers no longer have to fight as much with doctors, as doctors and insurers have a common financial interest. Second, "risk," the very basis of insurance, is offloaded on physicians. The physician bears the risks of funding expensive patients if they should happen to be in the physician's group.

4. CAPITATION'S INCENTIVES

To measure the strength of the incentive to reduce work, you need to look at the shape of the curve: $/unit versus units worked. For example, an increase in fee/unit by 50% for a 10% reduction in work is a greater incentive than an increase of only 10%. In figure 3 (Appendix A), we see the strong effect capitation has. Indeed, as physicians do less, the incentive to decrease care further becomes even stronger.

Though insurers initially claimed that the least expensive care was the best care, that view was not widely held. There is a consensus that there needs to be some decrease of care, to the point where low-value high-cost care is eliminated, but capitation is a poor attempt to address that specific problem.

> The formal reward system should positively reinforce behaviors, not constitute an obstacle to be overcome. –Steven Kerr[4]

Why capitation, if it does not properly address the problem? Kerr lists several reasons why reward systems may be developed that encourage behaviors other than the stated goals.[4] For insurers, at least two were operative. First, simplicity: it is very hard, if not impossible, at present to measure the value and quality of many medical services. Absent these measures, disbursing a fixed pot of money to the physician group for them to deal with is a simple way of disposing, if not dealing with, the measurement problem. Second hypocrisy: there is no doubt that the desire to pay less was the ultimate driver behind capitation. Quality of care was in no way safeguarded. By claiming that cheap care was quality care, insurers were able to cloak this decision in apparently well meaning garb.

5. ALTERNATIVES TO CAPITATION

If not capitation, then what? It is clear that we spend too much money on health care and that something must be done. Ideally, quality and high-value health care would be rewarded. This is certainly a goal everyone would support, at least in theory. However, the state of today's knowledge is too poor to be able to measure any but the most simplistic measures of quality, such as patient waiting time, immunization rates, etc.[5] Rewarding quality of care for more complex patient issues is impossible for now. The complexity and uniqueness of patients makes the goal of solely rewarding quality and value perhaps impossible to achieve.

We are left with a half-solution, and we should acknowledge that state of imperfection by avoiding strong incentives such as capitation, which can seriously harm the delivery of appropriate care.

> Virtue is a mean...between two vices, the one involving excess, the other deficiency. —Aristotle[6]

Designing an incentive that urges some reduction without strongly encouraging extreme decreases in care best reflects our present imperfect state. As Aristotle acknowledges, finding the virtuous mean can be extremely difficult. We do not now know where that point is.

Utilizing the knowledge of the effect of the payment/unit slope upon incentives, one can design a curve to match this goal. Figure 4 shows the payment/unit and the slope of the payment/unit of one such "designer curve". Here the incentive (slope) is greatest at levels of service similar to current or slightly less than current utilization levels. By replacing the capitation payment with a better design, the strong incentive to drive utilization to zero is eliminated.

Some may argue that physicians are not automata, and will not automatically follow incentives when the incentive would harm patients. Indeed, my proposed redesign of incentives is based on a simpler, behaviorist view of human behavior.

> As for responsibility and goodness—as commonly defined—no one...would want or need them. They refer to a man's behaving well despite the absence of positive reinforcement that is obviously sufficient to explain it. Where such reinforcement exists, 'no one needs goodness.' –G. E. Swanson[7]

Most ethicists would argue that there is more to human behavior than mere responses to reinforcement. But human nature is such that money does influence behavior, and not all physicians have such high altruistic impulses as to render ineffective the excess influence of unwisely designed incentives. Just because an insurer hands a physician group payment in the form of capitation does not mean it is either necessary or wise for the group to do the same with individual physicians.

ENDNOTES AND REFERENCES

1. PJ Feldstein, "Health Policy Issues: An Economic Perspective on Health Reform," Ch. 1 (Ann Arbor, Michigan: AUPHA Press), 1994.
2. Adam Smith, *The Wealth of Nations*, (NY: Collier), 1992.
3. Oscar Wilde, *The Picture of Dorian Gray*, (NY: Modern Library), 1992.
4. Steven Kerr, "The Folly of Rewarding A, while Hoping for B", *Academy of Management Journal*, 1975, 4: 769–783.
5. "Physician Performance: Report Cards under Development but Challenges Remain," United States General Accounting Office Report, GAO/HEHS-99-178, September 1999.
6. Aristotle, *Nicomachean Ethics*, tr. By WD Ross, Book II, Chapter 9, (London: Oxford University Press), 1925.
7. GE Swanson, Review symposium: "Beyond freedom and dignity," *American Journal of Sociology*, 1972, 78: 702–705.

APPENDIX A

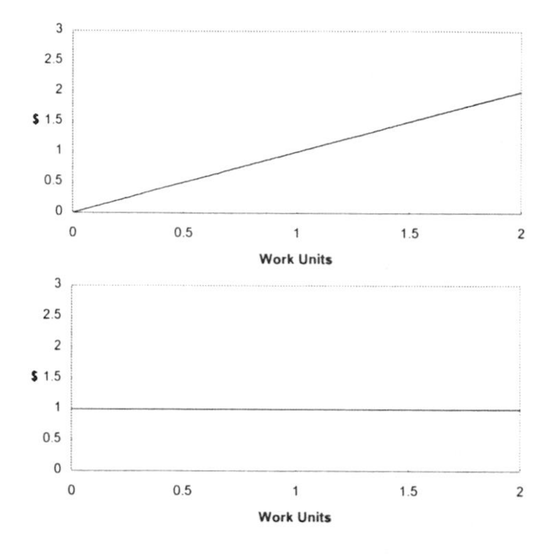

Figure 1. Total payment to physician group for taking care of a group of patients, comparing fee for service to capitated payment schemes. 1 work unit represents the typical amount of work done to care for those patients. $1 represents the amount of money received for capitated care for that group of patients.

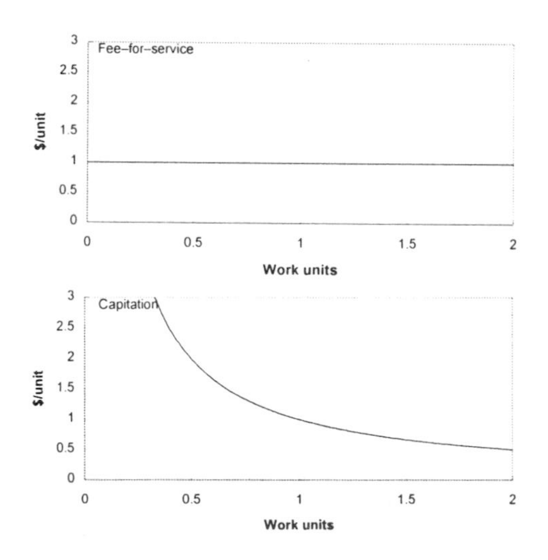

Figure 2. Payment per unit of work. Work units are as defined in figure 1. Capitation payment per work unit rises to high levels when work is reduced.

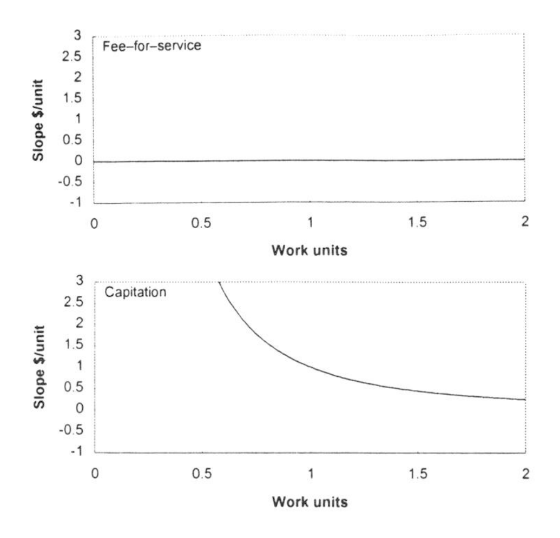

Figure 3. Strength of incentive (slope of payment per unit of work). Fee for service is constant payment per unit, so its slope is zero. The incentive of capitation rises to very high values at low work levels.

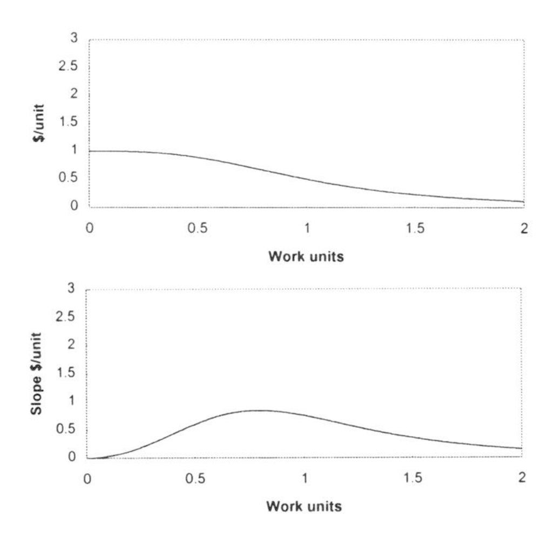

Figure 4. Example of a payment plan designed to maximize the incentive at utilization levels slightly below current level. The peak slope $/unit is slightly below normal utilization. The magnitude of the peak is also reduced to reduce the maximum effect of the financial incentive.

APPENDIX B

Assume that the usual payment is $1 per unit of work and that the normal amount of work for the group of patients is 1 work unit. Slope of payment is the derivative of the payment per unit of work (graphed as positive values for ease of comparison).

Payment type	Total payment	Payment per unit of work	Slope of payment per unit of work
Fee for service	$1 \times work\ units$	1	0
Capitation	1	$\frac{1}{work\ units}$	$\frac{-1}{work\ units^2}$
"Designer curve"	$\frac{work\ units}{1 + work\ units^3}$	$\frac{1}{1 + work\ units^3}$	$\frac{-3 \times work\ units^2}{(1 + work\ units^3)^2}$

A general form of the "designer curve" is:

$$\frac{target^{\lambda}}{target^{\lambda} + actual^{\lambda}}$$

where *target* is the amount of work units at which 50% of the maximum incentive is given and *actual* is the actual amount of work units performed. λ (lambda) governs the steepness of the curve (increasing λ increases steepness). Steepening the curve narrows the range of utilization in which there are significant changes in payment per work unit and increases the peak slope.

The general equation for the slope of the "designer curve" is:

$$\frac{-\lambda \times target^{\lambda} \times actual^{\lambda-1}}{(target^{\lambda} + actual^{\lambda})^2}$$

SPECIFIC APPLICATION IN UC DAVIS MEDICAL GROUP

For the UC Davis Medical Group a large portion of the capitation money is paid on a fee for service basis. The amount paid is based on historical data, and the goal is to have a relatively small amount left over at the end of each quarter for incentive payments. The surplus at quarter's end is then distributed to the departments based on each department's performance compared to its target. Mathematically, the payment for each service is:

$$\text{Initial payment: } \textit{work units} \times \textit{initial payment factor}$$

$$\text{Incentive payment: } \textit{work units} \times \textit{incentive factor} \times \frac{\textit{target}^3}{\textit{target}^3 + \textit{actual}^3}$$

$$\text{Total payment: } \textit{initial payment} + \textit{incentive payment}$$

where *work units* are the Medicare Fee Schedule RVUs (relative value units) for the service, *initial payment factor* is the dollars initially paid per RVU, and *incentive factor* is adjusted at the end of each quarter to be sufficient to distribute the entire surplus. The value of *incentive factor* is the same for every department. Most spreadsheet programs can easily determine the *incentive factor* through goal–seeking, and can then readily calculate the incentive distribution.

Chapter 14

Cross Cultural Issues in Medicine

An Integral Component of the Health Care System Plan

Jeanny K. Park
Assistant Professor of Clinical Pediatrics
University of California, Davis
Sacramento, California 95817
jkpark@ucdavis.edu

Key words: health care systems, cultural diversity, community sensitivity, access to health care, cost control, quality of care, decisionmaking analysis; cost, benefit, efficacy and outcome assessment

Abstract: With a population of 33 million people, Californians represent approximately 12 percent of the entire US population. The face of California, as well as all of America, has and will continue to become increasingly racially, ethnically, and culturally diverse. By the year 2010, California's white non-Latino population will no longer be the majority group. These facts highlight by sheer numbers the challenges that California health care systems face as they seek to provide excellent health care while competing for a culturally diverse patient base. Competing needs must be juggled—e.g., improving access to health care services, controlling costs, maintaining community and cultural sensitivity, demonstrating acceptable health care outcomes.

Everyday contradictions illustrate the difficulties in balancing these competing priorities. Though the Patient's Bill of Rights is espoused by every health care system, a hidden agenda reflects unspoken attitudes. That Bill of Rights only truly applies to those who can negotiate the cultural mores of mainstream American culture. For example, interpreting services is often viewed as an expendable service, yet length of stay, an important quality of care and financial marker, will be adversely affected when patients refuse to cooperate with care plans because of a culturally based misunderstanding. Similar to preventive medicine, directly addressing cross-cultural issues in medicine with an initial investment of time and money will improve outcomes. More importantly, the

Changing Health Care Systems from Ethical, Economic, and Cross Cultural Perspectives, edited by Loewy and Loewy. Kluwer Academic/Plenum Publishers, New York, 2001.

essence of providing health care is to truly listen to the patient, to understand and to care. The true mission of a health care system is to approach the community to meet their needs. To succeed in achieving these objectives, an organized effort to address cross-cultural issues must be an integral component of the strategic plan of any health care system.

1. INTRODUCTION

A 6-month-old baby girl is brought to the emergency room by her parents because she has a fever. Though the mother is bilingual and speaks both English and Spanish fluently, the father speaks Spanish only. The medical encounter is conducted predominantly in English between the medical student and the mother. In the emergency room the baby has no fever and a source for the reported fever cannot be identified. Throughout the encounter, the medical student makes repeated references to the necklace that the baby is wearing, telling the mother that it is dangerous and should be removed. The attending emergency room physician performs the final assessment and reassures the parents that the baby probably has a mild cold. As the medical student and the attending leave the room, the student exclaims, "Why is the baby still wearing that necklace? Didn't I tell you that it was dangerous?"

At the conclusion of this visit, both the parents and the health care providers are left with feelings of dissatisfaction, distrust, or worse. Excluded by language, the father feels insulted by the lack of respect accorded him as the leader of the family. He is still concerned about the health of his child. The mother feels upset by the implication that she is not a good mother. The medical student walks away frustrated and confused about how to make patients "comply." Perhaps, worst of all, the attending physician continues on with business as usual, not realizing that this family has decided to avoid coming to this hospital ever again.

Every one of us possesses a unique cultural background, which, often imperceptibly, affects our actions, expectations and assumptions: how we think, speak and act. "Culture" incorporates language, thoughts, communications, actions, customs, beliefs, values and institutions of racial, ethnic, religious or social groups."[1] As such, culture plays a central role in the delivery of health care, ranging from defining the parameters of health and illness, to determining who should provide treatment and what that treatment should be.

The face of America has and will continue to become increasingly racially, ethnically, and culturally diverse. Though this diversity fuels economic and social growth it also brings a new set of challenges to health care systems. This paper seeks to examine these challenges and to advocate for changes to successfully address these issues. First the rationale for integrating cultural competence into the health care system will be discussed. Next, the specific challenges engendered will be introduced. Finally, a review of recommendations to incorporate into the health care system strategic plan will be presented.

2. RATIONALE FOR INTEGRATING CULTURAL COMPETENCE

To those who already believe in the necessity of cultural competency in any arena, a dialogue on cultural awareness is often an exercise in "preaching to the choir." However, for many, the pervasive effects of cultural differences on health care delivery are often downplayed or even completely ignored. Thus, a rigorous review of the rationale for integrating cultural competence into the health care system serves to meet the important challenge of educating and persuading those who can effect change, whether they be policy makers, legislative bodies, health system administrators, or health care providers.

Reviewing the mission statements of local health care systems provides a logical starting point. Stated goals explicitly outlining the organization's purpose are characterized by common themes including:

1. *Provision of quality health care*
    - Enhance the health and well being of people in the communities we serve through compassion and excellence
    - Create a quality, patient-centered, integrated provider network
    - Satisfy the health care needs of our members
    - Place special emphasis on providing care to vulnerable populations within our society
2. *Maintenance of financial viability or profitability*
    - Operate as a unified health system that can compete in a managed care environment
3. *Education*
    - Provide a primary clinical site[2, 3, 4]

Immediately questions arise. "Who" is considered "the community?" Have the health care needs of an increasingly diverse patient population been identified? What constitutes "quality" health care, particularly as this relates to culturally defined expectations? What about legitimate economic concerns regarding costs, financial stability, capital formation and growth, competition for market share, and profitability? Does the health care system's objective to educate extend to the community, policy-making organizations and legislative bodies?

First let us examine who lives in the communities we serve. With approximately 33 million people (approximately 12% of the entire population of the United States), California serves as a bellwether for national trends. In 1996, 53% of the California population was categorized as White, non-Hispanic, with the remaining 47% comprised of Hispanic (29%), Asian (10%), African American (7%) and Native American (1%) populations. By 2010, the population of California will grow to 41 million people. Because of higher growth rates in the Hispanic and Asian populations from births, immigration and longer life spans, the White, non-Hispanic population will decrease to less than 50% of the state's population. No single racial or ethnic group will constitute a majority population.[5]

Numerically, diversity has become the norm. In 1990, 32 million people, or 14% of the US population, reported speaking a language other than English in the home. In fact, it is reported that 328 different languages besides English is spoken in the United States.[6]

What about the quality of the health care provided? Despite differences in culturally defined expectations, life expectancy represents one readily understood measurement. It is distressing to find that despite advances in medial technology, the average life expectancy of African Americans in California is significantly less than that experienced by the White population.[5,6,7] This finding has largely been attributed to the higher death rates from cardiovascular disease and cancer experienced by the African American community.[8]

Examination of difference in death rates for diabetes by race and age further illustrates the complexities of providing quality health care to a diverse population.[9] African American and Hispanic populations suffer from disproportionately higher death rates caused by diabetes in comparison to the White population. What factors contribute to these findings? Are they related to differences in disease burden or natural history, socioeconomic status, access to or quality of health care? What role does racial or ethnic background play? What is the cost of providing health care to these populations? What can be done to improve these outcomes, and presumably enhance cost effective care?

The data and the questions they engender highlight the conflict between the "Art" and the "Science" of medicine. Most health care providers believe or behave in a fashion that places the science of medicine above culture. If the provider can diagnose the disease and institute the evidence based medical treatment, success will follow. However, as the initial scenario demonstrated, the reality is that health care is a cultural construct. In a manner of speaking, the Patient's Bill of Rights summarizes the "art" of medicine by delineating the patient's right to considerate and respectful care, to information (regarding health, financial, and medical research), to privacy and confidentiality, and to autonomy in consenting to or declining treatment.[10] Failure to recognize and incorporate this perspective in a culturally sensitive manner ignores the moral imperative outlined in the mission statements, and will lead to failure to improve the health of a progressively increasing proportion of the population.

From an economic perspective, the importance of maintaining the health and thus economic productivity of an increasingly diverse community is obvious. The more focused viewpoint addresses diversity as a business imperative. Particularly in a managed care environment, competition for a capitated patient base leads to the targeting of growing minority populations for enrollment and maximization of retention rates.[11] The business sector has already acknowledged the increasing buying power of minority populations in their marketing plans and advertising strategies. Finally, the data above demonstrates that analyzing the differing health needs of a diverse patient population may enhance the use of tools such as evidence based medicine, benchmarking and critical pathways to achieve more effective cost containment.

A growing body of legal mandates regarding cultural competency is a final and increasingly important issue. At the Federal level, Title VI of the Civil Rights Act of 1964 states: "No person in the United States shall, on ground of race, color or national origin, be excluded from participation in, be denied the benefits of, or be subjected to discrimination under any program or activity receiving *Federal financial assistance*."[12] Most health care systems fall under this mandate as they care for Medicare/Medicaid participants. Similarly, the Hill Burton Act of 1946 provided construction funding for many hospitals with the provision of a "community service obligation" for recipients lasting in perpetuity.[13] As noted above, the communities served today are characterized by their diversity.

State laws regarding cultural or linguistic considerations are quite variable, but address issues governing language access, state civil rights, state managed care contracts and medical malpractice. From the perspective of tort law, state statutes primarily affect language obligations for patients with limited English proficiency. Providers may be liable for absence of informed consent, breach of professional standard of care through failure to communicate or a presumption of negligence.[14] The importance of this issue is recognized by the Mutual Insurance Corporation of America which offers a discount on malpractice insurance to those physicians who participate in cultural competence training.[15]

Accreditation agencies such as the Joint Commission on Accreditation of Healthcare Organizations (JCAHO) and the National Committee on Quality Assurance (NCQA) have developed standards primarily regarding provision of translation services.[16,17] In response to the hodgepodge nature of these legal and regulatory mandates, the U.S. Department of Health and Human Services Office of Minority Health has organized an initiative to develop a national consensus on this issue and to draft national standard language.[18] Currently undergoing a process of public comment to encourage dialogue and final revision, these comprehensive standards seek to provide "empirically justifiable and practically viable" recommendations.

Thus, the rationale for integrating cross-cultural competence into the health care system plan recognizes the moral imperative to fulfill explicitly stated mission goals, the economic perspective to maximize cost effective behavior and the mandates of legal and regulatory standards at the national and state levels.

3. RECOMMENDATIONS FOR AND CHALLENGES TO IMPLEMENTATION OF CULTURAL COMPETENCY

The recent development of draft national standards for assuring cultural competence in health care synthesizes what has been to date a piecemeal collection of information and recommendations. The standards serve as a blueprint of recommendations to be implemented by health care systems.

The first standard asserts the need for health care organizations to promote and support *the beliefs and attitudes* that cultural and linguistic competence are fundamental and integral to providing health care services. To accomplish this, the health care system must recognize the diversity of the community it serves and endorse this belief with an explicit and comprehensive organizational commitment. Specifically incorporating a commitment to cultural competency or sensitivity into the institution's mission statement and designating a task force or work group to develop written policies force the issue into the open and provide the door to change. Other recommendations for change include, but are not limited to:

1. Expanding health insurance coverage
2. Improving access to primary care and preventive services
3. Implementing community needs assessment
4. Collecting and utilizing accurate outcome data by cultural group
5. Providing culturally and linguistically appropriate services

6. Initiating outreach to local and cultural communities
7. Evaluating progress, cost and benefits
8. Recruiting, educating and training a diverse group of administrative, clinical and support staff

The dilemma in implementing these recommendations is obvious. How much will it cost? Who pays? Which interventions are the most effective? How do you perform self-assessment and ensure performance improvement? What analytical tools exist to produce data sufficient to analyze the competing interests of financial incentives versus the needs and expectations of individual patients or populations?

Language barriers have received the most attention and analysis as obvious obstacles to providing culturally competent care. Providing those with limited English proficiency access to bilingual staff or interpretation services obviously decreases the chance for "miscommunication, misdiagnosis, inappropriate treatment, reduced patient comprehension and compliance, clinical inefficiency, decreased patient and provider satisfaction, malpractice, injury, and death."[18] At the single provider level, providing comprehensive language assistance may not be a feasible option. However, at an institutional level, lack of implementation may be due to ignorance of legal obligations, inadequate know how or cost considerations. One framework for integrating a language assistance program includes:

1. *Linking* information regarding the language needs and preferences of patients to available resources to insure access and continuity of care
2. *Checking* for uniform quality, which requires training for both interpreters and provider, establishes standards
3. *Tracking* costs and utilization assists in measuring and improving services.[19]

What are the costs for providing language assistance and who pays for language services? Few formal studies have been performed to determine costs. Informal analysis of interpreter services performed at six selected health care facilities demonstrated a wide variance in total budgets. Variability in patient population served, methods of providing language assistance, training programs and so on all contribute to this variability. On the other side of the coin, interpreter services can be recognized as a billable service for Medicaid reimbursement.[20] Some managed care companies have taken advantage of their organizational strengths to deliberately plan and implement cost effective delivery of language services, enabling them to acquire further market share through targeted enrollments or enhance their attractiveness to employers and other purchasers of health care services.

4. CONCLUSION

Health care systems will find that cultural diversity in the communities they serve is the rule and not the exception. Compelling reasons—moral, economic and legal—exist to explicitly incorporate cultural and linguistic competency as essential elements of the health care system strategic plan. The recent emergence of draft national standards for cultural and linguistic competency provides a timely blueprint for implementing institutional change. The usual challenge of juggling competing demands for time, money and commitment are recognized. However, it is simplistic to assume that the disturbing discrepancies in health care outcomes among different racial groups can be completely ascribed to financial constraints. Further data is required to assess costs versus benefits, and to measure efficacy and outcomes. Development of a research agenda to assist in decision-making analysis must become a priority.

ENDNOTES AND REFERENCES

1. Office of Women and Minority Health, Bureau of Primary Health Care, HRSA.
2. Sutter Health (www.sutterhealth.org) "Mission, Vision, Values," 1999.
3. Kaiser Permanente California (www.kaiserpermanente.org) "Our Community Service," 1999.
4. UC Regents (www.ucdmc.ucdavis.edu) UC Davis Health System, 1999.
5. State of California, *California Health Care Fact Book Sacramento: Office of Statewide Health Planning and Development*, 1999.
6. U.S. Bureau of the Census, *Statistical Abstract of the United States*, Table 57 (Washington, DC: Government Printing Office), 1996.
7. National Center for Health Statistics, "Health, United States, 1990," (Hyattsville, MD: U.S. Public Health Service), 1991.
8. U.S. Department of Health and Human Services, "Report of the Secretary's Task Force on Black and Minority Health," vol. 1, (Washington, DC: DHHS), 1985.
9. National Center for Health Statistics, "Health, United States, 1997," (Hyattsville, MD: U.S. Public Health Service), 1997.
10. American Hospital Association, "A Patient's Bill of Rights," (Chicago, IL), 1992.
11. Manfiletto E. Profit in diversity. Managed HealthCare 1995 Jul:36-46.
12. Civil Rights, Subchapter V, Federally assisted programs. U.S. Code. Vol. 42, Sec. 200d, 1992.
13. "Construction and Modernization of Hospitals and other Medical Facilities," U.S. Code. Vol. 42. Sec.291, 1994.

14. J. Perkins, H. Simon, F. Cheng, K. Olson, Y. Vera, "Ensuring Linguistic Access in Health Care Settings: Legal Rights and Responsibilities" (Palo Alto: The Henry J. Kaiser Foundation), April 1998.
15. RS Trosty, Presentation at the Quality Health Care for Culturally Diverse Populations conference, New York, October 1-4, 1998.
16. Joint Commission on Accreditation of Healthcare Organizations, "Comprehensive Accreditation Manual for Hospitals," 1997.
17. National Committee for Quality Assurance, "Availability of Language Interpretation Services: Summary of Changes from HEDIS 2.5 and/or Medicaid HEDIS, 3.0," vol 2, 1997.
18. JP Fortier, *et al*, "Assuring Cultural Competence in Health Care: Recommendations for National Standards and an Outcome-focused Research Agenda," Federal Register, 1999.
19. PH Chang and JP Fortier, "Language Barriers to Health Care: An Overview," *Journal of Health Care for the Poor and Underserved,* 1998; 9 (supp.): s5-s20.
20. J. Hornberger, "Evaluating the Costs of Bridging Language Barriers in Health Care," *Journal of Health Care for the Poor and Underserved,* 1998; 9 (supp.): s26-s39.

Chapter 15

Competing Interests in Pediatric Managed Care Settings

Susan E. Zinner-Kemp
Assistant Professor
School of Public and Environmental Affairs
Indiana University
Gary, Indiana 46408
e-mail: szinner@iun.edu

Key words: health care systems, managed care, patient vulnerability, goals of medicine, physician responsibility, competing interests, public policy, allocation of resources, access to health care

Abstract: How well do managed care organizations meet the health care needs of children? What guides administrators and physicians as they participate in helping to make medical decisions involving children? Perhaps, more importantly, what *should* guide them? Are managed care organizations better at meeting the medical needs of some children and not others? If so, who and why? I intend to explore how medical decisions can be made that are in the "best interests" of both the insured child and the MCO. This entails a discussion of communitarian justice—regarding allocation of resources and rationing decisions—as well as attention to the acute health care needs of children as unique individuals with unique problems. What is the ethically appropriate response from MCOs to these patients? Is a utilitarian framework too harsh? Is a deontological model too expensive? My goal is to stimulate the development of some fundamental guidelines to assist MCO administrators, physicians and others when addressing the medical needs of all children.

Managed care organizations (MCOs) were created in an attempt to provide quality medical care to the American public while saving health care dollars in an era of rapidly-escalating costs. Despite several well-publicized

Changing Health Care Systems from Ethical, Economic, and Cross Cultural Perspectives, edited by Loewy and Loewy. Kluwer Academic/Plenum Publishers, New York, 2001.

failures to achieve this goal, many feel that this goal has been largely attained. What is less clear, however, is whether the needs of children have been overlooked in the process.

Childhood health needs and services differ significantly from those of adults. Children need preventive services; acute services for illnesses and injuries; management of developmental, school, psychosocial and emotional problems; and the occasional use of emergency and inpatient care[1] Adults, on the other hand, require a system which emphasizes acute episodic care and chronic disease management.[2]

The child's status also poses a potential problem in accessing health services. Children are uniquely dependent for all of their needs, including health care, on adults due to their cognitive and emotional vulnerability.[2] Childhood is also a crucial time as the patterns of health care established during this time may determine lifelong habits which impact health status later in life[2]. These factors make the provision of health services during childhood essential.

In theory, MCOs are well positioned to meet the needs of children. MCOs offer preventive services designed to cure or prevent illnesses or conditions early in their development and thereby reduce or eliminate the need for future, more complex and costly treatments. MCOs tend to have more data on their covered patients than traditional fee-for-service providers and emphasize primary care, prevention and service coordination.[1] On the other hand, MCOs tend to limit access to specialty services and prefer to provide referrals to specialty services only when a demonstrable benefit is anticipated.[3] The American Academy of Pediatrics' (AAP) Committee on Child Health Financing was sufficiently concerned about the potential negative ramifications of MCOs on child health to create guiding principles for MCOs providing services to children.[4] The committee lists five such principles of concern.

The first of these principles requires that children should have access to appropriate primary care providers This principle encompasses having a primary care pediatrician available at all times, educating the family about MCO rules and operation and the need for the primary care pediatrician to act as gatekeeper.[4] The second principle involves access to pediatric specialty services. This principle includes elimination of barriers to access and ensuring that an appropriate mix of specialists exists in each geographic area of the country. Also significant is the need to establish referral criteria.[4]The third principle involves treatment authorization and urges parents to be aware of the MCO's participating providers and ensuring that a process for timely responses to both treatment authorization requests and an appeals process exists. The fourth principle focuses on the development and utilization of quality assurance mechanisms within the MCO. The final principle considers the roles of financing and reimbursement and suggests that capitation rates be developed to cover all of the suggested AAP preventive services through age 21.[4]

These guidelines remain merely guidelines. Their implementation would significantly reduce a number of problems currently facing MCOs providing services to children. However, one issue facing pediatricians is that MCOs were created to meet the needs of the majority of patients seeking care, including children. While a utilitarian approach is laudable and does serve to benefit the majority, those in special circumstances may find themselves facing insurmountable financial and regulatory obstacles.

Chronically ill children are one group whose needs often remain unmet. Their needs include coordination of care, continual assessment of both child and family development and having a provider who serves as an advocate for the child and family.[5] One survey found that MCOs have policies that often prevent chronically ill children from receiving necessary services.[3] Some MCOs provide specialty services only when significant improvement is likely within a short period, place limits on the amount and duration of services frequently needed by chronically-ill children and limit both choice and access to specialty providers. Further, the survey authors found that "to a large extent, the availability and quality of services available to a child with special needs is likely to depend on the parents' ability to maneuver within the system."[3] This imposes a great burden on parents and guardians who must be willing and skilled enough to challenge administrators, physicians and bureaucracy.

Similarly, children with disabilities face a number of obstacles. While this group consumes a disproportionate amount of services, they continue to have unmet needs. Michigan's attempt to enroll disabled children in its Medicaid managed care program identified problems such as inadequate reimbursement, lack of knowledge of existing resources, poor communication between providers, lack of interested gatekeepers and time constraints.[6] MCOs are structured to address the needs of children with acute illnesses. While most children are likely to benefit from care provided by MCO providers, children facing chronic illness and disabilities face a system that is likely to fail them. How should we respond to this group? Is a utilitarian framework too harsh for children with special needs? Is a deontological model too expensive?

This is one of several competing interests that pit MCO patients against other MCO patients. A second set of competing interests involves disincentives for providers, generally primary care pediatricians, to refer patients for tertiary services. The relationship between a physician and patient is fiduciary in nature. That is, due to the inequity in the relationship, doctors are held to a high standard in their relationships with patients. An MCO arrangement that creates incentives for physicians not to refer patients threatens physicians' fiduciary obligations to patients.

Rodwin has noted that physicians can play three different roles in this context. They can act as "ideal fiduciaries" by promoting their patients' interests without regard to competing obligations, they can act as "neutral resource allocators" and they can act in their own best financial interests or in the interest of a third party such as the MCO.[7] A survey designed to test which of these three standards was most common studies the relationship between financial incentives and hospitalization rates and outpatient visit rates.[8] A study of 302 HMOs found that the use of capitation or salaries as payment mechanisms was associated with lower rates of hospitalization and fewer outpatient visits than in traditional fee-for-service arrangements.[9] As Bergman and Homer note:

> (A)lthough this study suggests that financial arrangements do affect physician decision making, there is a paucity of evidence to show that the decreased utilization occurring in HMOs results in worse patient care outcomes for children.[8]

Whether the physician stands to benefit from lower expenditures in an MCO setting or experiences pressure to save dollars in a hospital-based DRG environment, physicians are likely to feel significant pressure to keep costs low.[10]

Interestingly, however, it has been suggested that just the opposite is occurring. A February 1999 *New York Times* article provides preliminary evidence that "fears of consumer backlash, legislative intervention or large jury awards" may be encouraging MCOs to provide nearly all requested services and treatments, whether medically appropriate or not.[11] Appeals from patients after denial of treatment have been very low and may jeopardize anticipated cost savings.[11] The evidence remains limited, however. An important question, therefore, is how to create a system that allows a physician to operate within an MCO framework and still meet the needs of patients. Perhaps more important, how can physicians develop the administrative and ethical skills to accomplish this task?

One set of competing interests—those of well children and children with special needs—involves the need to allocate fairly within an MCO and at the societal level. The second set of competing interests, those of children and their physicians who fail to act in their best interests, involves the need resolve the tension between the MCO's financial mechanisms and the physicians' fiduciary obligations. This issue also involves resource allocation of a more limited nature; that is, should health care dollars be devoted to patient care or should all or a percentage of those dollars reward physicians who limit care (whether appropriately or inappropriately)?

Daniels and Sabin have proffered four MCO rationing guidelines:

1. Decisions regarding coverage for new technologies (and other limit-setting decisions) and their rationales must be publicly accessible
2. Rationales for coverage decisions should aim to provide a reasonable construal of how the organization should provide "value for money" in meeting the varied health needs of a defined population under reasonable resource constraints (specifically, a construal will be "reasonable" if it appeals to reasons and principles that are accepted as relevant by people who are disposed to finding terms of cooperation that are mutually justifiable)
3. There is a mechanism for challenge and dispute resolution regarding limit-setting decisions, including the opportunity for revising decisions in light of further evidence or arguments
4. There is either voluntary or public regulation of the process to ensure that conditions 1 through 3 are met[12]

The authors note that the focus of these four principles is to move the debate from the confines of the MCO into a public policy debate. Ensuring a societal discussion will help us decide "how to use limited resources to protect fairly the health of a population with varied needs, a problem made progressively more difficult by the successes of medical science and technology."[12]

Exploring these issues as a society is an important first step in this debate. Rationing decisions are currently made within the MCO and the fact that these decisions are not made in public and that the MCO may be unable to satisfactorily articulate a justification for the decision to deny treatment concerns many health care workers and patients. Creating an environment where MCOs can publicly establish the legitimacy of their decisions based on previously established principles would go a long way to resolve these debates. In fact, MCOs do weigh competing values and make decisions each day.[12] By opening up the debate to include comments from the public, we can explore the appropriateness of those decisions. We can also begin a debate about the scale used by the MCO. Are the values used important ones? Should we amend the scale at all to give more or less weight to certain values? If so, which ones?

One writer has noted that the delicate balance between individuals and populations is similar to the issues facing physicians distributing the polio vaccine, where small risks are deemed socially acceptable if the larger community benefits. He notes that "(t)he issue is not that we have to weigh both the individual and the general good, but that we do it with the right reasons in mind: for the promotion of overall health and not for individual profit."[13]

Given the tension inherent in allocation decisions, the following guidelines should be considered.

A. Establish higher capitation rates for disabled and chronically ill children.

The relatively small number of children with disabilities or who are chronically ill consume about 90 percent of the health care dollars spent on pediatric services.[14] MCOs have no incentive to enroll or provide medically necessary services for this population. In fact, "until the plans have a large number of children with special needs, and the capitation rates they receive reflect these children's additional unique needs, HMOs will have an incentive no to enroll them, or to undertreat children with special needs that they are forced to enroll."[6] Changing the capitation rates for this small group of children would have a relatively insignificant impact on U.S. health care costs and yet may create incentives for MCOs to encourage provision of medically necessary services to disabled and chronically ill children.

B. Restructure MCO financial incentives so pediatricians are encouraged to provide all medically necessary referrals and services.

Instead of financial incentives designed to limit referrals to specialists and other expensive services, physician financial incentives could be changed to promote referrals when medically appropriate, including out-of-plan referrals. This will, in turn, create an obligation to ascertain the needs of patients. Provides will be held accountable for the quality of the care they deliver and the resources they utilize. Using this approach offers the ethical advantage of "link(ing)...[the financial incentive] to the correctness of clinical decisions, not general volume targets.[7]

C. Continue to focus on preventive services for the general population of well children.

MCOs should continue their efforts in what remains their greatest success, i.e., providing services to the majority of the pediatric population in need of routine well child care with occasional acute needs. Recent studies have found higher pediatric immunization levels, more screening tests, more frequent and complete checkup visits and more visits for allergy care and other treatments for children enrolled in MCOs compared to traditional fee-for-service coverage (Balaban *et al.*, 1994; Godrey & Christiansen, 1995; Mustin *et al.*. 1994' Szilagyi *et al.,* 1992; Valdez *et al.*, 1989, cited in Leatherman & McCarthy, 1997).[15]

D. Move the debate about allocation and rationing decisions from the MCO to the public.

The Daniels and Sabin guidelines discussed earlier provide a good starting point. Comments from parents, patients, administrators and providers help establish a legitimacy to MCO decisions that is currently missing. When we let all players make their values and concerns public, we have the beginning of an important public policy debate.

While the ethical principle of distributive justice does not tell us how to resolve these competing values, it will provide some guidance.[12] With the likely result of MCOs assigning different weights to values, the likelihood of different decisions occurring in different MCOs is inevitable. Does this raise the issue of relativism in decision-making or is this an acceptable outcome in a pluralistic environment? According to Daniels and Sabin:

> If we as a society can tolerate the inevitable differences in decisions and policies that the different configurations of values will create, we will have an opportunity to learn from the dialectic between principle and practice. We will see more clearly through a legacy of specific decisions and the outcomes just what the moral and nonmoral benefits and costs of different approaches are.[12]

In other words, MCOs can learn from decisions made by their counterparts and consider whether the weights assigned to specific values should be changed.

In conclusion, the growth of MCOs need not endanger the health status of children with special needs. Both this at-risk population and the larger population of healthy children stand to benefit when their provider works for an MCO. These guidelines were created to serve the needs of children and still ensure that the MCO remains financially viable. In an era of cost constraints and shrinking health care budgets, we must focus on the best interests of our future: the American children.

ENDNOTES AND REFERENCES

1. PG Szilagyi, "Managed Care for Children: Effect on Access to Care and Utilization of Health Services," *The Future of Children: Children and Managed Health Care*, ed. by RE Behrman (Los Altos, CA: The David and Lucille Packard Foundation, 1998), pp. 39-59.
2. REK Stein, "Changing the Lens: Why Focus on Children's Health?" *Health Care for Children: What's Right, What's Wrong, What's Next?* Ed. by REK Stein (NY: United Hospital Fund, 1997), pp. 1-11.
3. HB Fox, LB Wicks and PW Newacheck, "Health Maintenance Organizations and Children with Special Health Needs," *American Journal of Diseases*, (1993): 147, 546-52.

4. "Guiding Principles for Managed Care Arrangements for the Health Care of Infants, Children, Adolescents, and Young Adults," *Pediatrics*, 95 (4), p. 613-15.
5. AP Eaton, KL Coury and RA Kern, "The Roles of Professionals and Institutions," *Caring for Children with Chronic Illness: Issues and Strategies*, ed. by REK Stein. (NY: United Hospital Fund, 1989), pp. 75-86.
6. DJ Lipson, and AB Bernstein, "Doing the Right Thing: The Role of Market Forces and Public Policy in Managed Care Organizations' Performance on Child Health," *Health Care for Children: What's Right, What's Wrong, What's Next?* (NY: United Hospital Fund, 1997), pp. 237-60.
7. M. Rodwin, *Medicine, Money and Morals: Physicians' Conflicts of Interest* (NY: Oxford University Press, 1993.
8. DA Bergman, and CJ Homer, "Managed Care and the Quality of Children's Health Services," *The Future of Children: Children and Managed Health Care*, ed. by RE Behrman (Los Altos, CA: The David and Lucille Packard Foundation, 1998), pp. 60-75.
9. AL Hillman, MV Pauly, and JJ Kerstein, "How do Financial Incentives Affect Physicians' Clinical Decisions and the Financial Performance of Health Maintenance Organizations?" *New England Journal of Medicine* 1989: 321, 86-92.
10. AR Fleischman, "Ethical Views and Values," *Caring for Children with Chronic Illness: Issues and Strategies*, ed. by REK Stein, (NY: United Hospital Fund, 1989), pp. 87-100.
11. MM Weinstein, "Managed Care's Other Problem: It's Not What You Think," *New York Times*, February 28, 1999.
12. N. Daniels and JE Sabin, "Last Chance Therapies and Managed Care: Pluralism, Fair Procedures, and Legitimacy," *The Hastings Center Report*, 1998: 28(2), 27-41.
13. JA Finkelstein, "Commentary: Defining the Challenge and Opportunities for Children in Managed Health Care," *The Future of Children: Children and Managed Care*, ed. by RE Behrman (Los Altos, CA: The David and Lucille Packard Foundation, 1998), pp. 138-40.
14. LW Deal, PH Shiono and RE Behrman, "Children and Managed Care: Analysis and Recommendations," *The Future of Children: Children and Managed Health Care*, ed. by RE Behrman (Los Altos, CA: The David and Lucille Packard Foundation, 1998), pp. 4-24.
15. S. Leatherman and D. McCarthy "Opportunities and Challenges for Promoting Children's Health in Managed Care Organizations," *Health Care for Children: What's Right, What's Wrong, What's Next?* (NY: United Hospital Fund, 1997), pp. 199-236.

Index